Helping your child with...

Home

Tests

For Grammar and Comprehensive School Entrance (Fourth Edition)

R.M. Arkwright

Harrap

First published in Great Britain 1956
by Harrap Limited, 19/23 Ludgate Hill, London EC4M 7PD

Reprinted: 1956 *(twice)*; 1957 *(twice)*; 1958; 1959; 1960; 1961; 1962; 1963; 1964; 1965; 1966; 1967
Second Edition 1968, *Third Edition* 1970
Reprinted: 1972; 1974; 1976; 1978; 1979
Fourth Edition: 1981
Reprinted: 1982 *(twice)*

ISBN 0 245-53786-4

Typeset by Quadraset Ltd, Radstock, Bath
Printed by Dotesios, Bradford-on-Avon, Wiltshire

Contents

Helping your child with Selection Tests

You will probably have bought this book because you want to help your child, who will shortly be taking a selection test of some kind. You naturally want your child to do as well as possible. How can you best help? Some of the advice given here may seem obvious, whilst some may be difficult to put into practice. However, the following suggestions will enable you to help your child do his or her best when they are eventually faced with a test.

Firstly, remember that most of the ground is already being covered in school and that anything you do should supplement, and not contradict or confuse, this. Since you left school a large number of things have changed. It is quite likely that you will find your child doing arithmetic in a different way from the methods you used. It would be most unwise to try and impose your methods on the child because you think they are better. This will only lead to confusion and it is very important that this is avoided.

Secondly, remember that your child puts in a full day at school and may have homework set, too. Do not pile on so much extra work that the child is overwhelmed and gives up trying. Take an interest in the child's school work in general and help if possible. However, resist doing the work for him or her.

Thirdly, guard against making your child over anxious. Exhibit interest by all means but do not let it show that you are unduly concerned about the results. If you do you may make your child over anxious and tense, and this may do great harm. Some children are naturally over anxious and worry anyway; they need careful and tactful handling to reduce their worry and give them confidence.

Many children, however, are far more happy-go-lucky. Your interest in them and their work can settle them down to it, but again too much pressure on your part may have a detrimental effect. The way you handle your child is most important for his or her success.

It is important that all children should know the fundamentals of English and arithmetic. Schools are occupied with this; you can help with your interest in, and advice on, homework but there are two ways in which you can give very positive help. The first is with tables. Find out just which tables your child is expected to know and then make sure they are thoroughly familiar to the child. The second is to rouse your child's curiosity in things and then to help him or her to satisfy it by finding things out independently using both his/her powers of observation and reasoning, and through the use of books. Make your child aware of what is going on in the world. Enthusiasm for knowledge is very important and you can help by providing the stimulus your child needs to develop it. By talking to your child about all sorts of general matters and giving him or her the chance to experience new places and things, you will help to develop a wide vocabulary which will help him or her in English. Encouraging reading will, of course, help this too. There are plenty of lively childrens' books written by contemporary authors. A dictionary to look up new words will help too.

This informal sort of help is perhaps more important than any other, but you should find this book useful in supplementing that help.

About this Book

Tests like these are used to select children for places in grammar schools and to sort children into streams after they have entered comprehensive schools. Experience in their use is therefore valuable to all children passing from primary to secondary schools.

Whatever the purpose is, the tests are usually in English and arithmetic or mathematics, and there is an intelligence test. The English tests normally include passages for the child to read and to answer questions on. There are usually also questions that test spelling and punctuation, possibly a little simple grammar, and a variety of tests that probe your child's knowledge of words, and his or her ability to use them. There may also be an essay or composition. The changeover to metric measurements and the rapid change from arithmetic to mathematics in junior schools have all been taken into consideration in designing the arithmetic and mathematics tests in this book. Schools are moving at different speeds, and some of the tests may seem strange. All should be tried, but if some cannot be managed there is no need to worry, for tests are unlikely to be set that are out of step with the work the children are doing.

The intelligence test is the most unusual to the child. It tries out his reasoning powers by asking him to work problems involving words, figures, and shapes. All the questions have the same end in view—to see whether the child can reason well. Although this test has its place among the others, it is only one of the tests used, and the others are just as important.

The basic types of test have been emphasized in this book. While working the tests the child will get used to the shape and form that they are likely to take in his or her examination, and therefore the examination will not come to him or her as something strange. Working through them will also help him or her to get used to jumping from one type of work to another, and to following instructions. This is most important. The instructions must be read and understood before any section of a test is started.

Most of the tests are rather shorter than they are likely to be in an examination. This enables them to be done in an evening without taking up too much of the child's time.

Children are expected to work these tests quickly, and for each a time limit is given. It is not expected that children should finish all the questions in this time. Rush may spoil everything. It is better to go along steadily and to get the answers right. For example, in the problem arithmetic tests encourage your child to do some rough working on a piece of paper rather than let him try to do all the working in his head. He will get more right in that way.

It is sometimes suggested that it is best to run through a test, doing all the easy questions first. Test questions are arranged in order of difficulty. Generally speaking, it is better to work steadily through, only missing a question if it seems to be taking too much time. A little more thought spent on the earlier and easier questions may be more rewarding than a hasty decision upon a later and more difficult one.

When tests have been worked teach your child to check his or her answers, then argue out any points that are not fully understood. Some of the problems may have answers that you think are just as good as the ones given here. If you feel this then call your answers right. Where gaps are found in the child's knowledge it may be possible to fill them.

If you have followed the advice given here you will have given your child a chance to do his best. All you can do after this is to wish him luck and keep your fingers crossed.

English Test 1 *(15 minutes)*

Read the passage carefully, then read each question given below and select the right answer from those given in the brackets.

As the boat entered the bay the wind dropped, so Tom took the sails down, and Fred picked up the oars and rowed in to the shore. The keel hardly touched the shingle beach before both boys leaped over the side and pulled it up safely above the high-tide mark. They then took their torches and raced up the beach towards the cave. Once inside they began to move more cautiously as though they expected trouble.

1. Why did Tom take down the sails?
   (Fred liked rowing : the boat was going too fast
   the wind had dropped : the sails were too heavy)

2. Who rowed the boat in?
   (Tom : a man : Fred : Robert)

3. What was the beach made of?
   (sand : rocks : mud : shingle)

4. How far up the beach did they pull the boat?
   (as high as they could : just out of the water : above high-tide mark : half-way out)

5. Why did they take torches with them?
   (because it was night : they were going into a cave
   they were frightened of the dark : they had got them with them)

6. How did they go up the beach?
   (running : walking : strolling : creeping)

7. How did they move in the cave?
   (cleverly : swiftly : slowly : cautiously)

8. What did they expect to find in the cave?
   (treasure : trouble : a man : a guide)

Robin planned carefully how he would decorate the dining-room for Christmas. Then he took a roll of coloured crêpe paper and cut it right through the roll into short pieces. He pulled out the middle of each piece without unrolling it until it formed a long twisted strip that he hung from the picture rail to the middle of the ceiling. From these he hung stars that he cut from gold and silver paper. Then he stretched string diagonally down each wall on which he hung all the Christmas cards he and his sister Mary had collected.

9. What did Robin do first?
   (took a roll of coloured paper : hung up stars : decided how he would do it : cut the coloured paper)

10. Who helped him collect the Christmas cards?
   (his father : his mother : his brother : his sister)

11. Where did he get the stars from?
   (he bought them : he cut them out : he borrowed them : he sent for them)

12. To what did he fasten the lengths of crêpe paper?
   (the ceiling : the lamp : the pictures : the floor)

13. How did he twist the paper?
   (with his fingers : round and round : he pulled out the middle of a roll : he unrolled them)

14. What colour were the stars he hung up?
   (red and green : gold and blue : silver and gold yellow and red)

15. Which room was Robin decorating?
   (kitchen : dining-room : bedroom : lounge)

16. What did he hang on strings?
   (stars : cards : rolls of paper : lights)

To reach the foot of Mount Everest the expedition had to march for days through deep, forested valleys. The long file of porters following each other set out at dawn each day, each man carrying a load of 50 kg. on his back. At ten in the morning the climbers set out to follow them. Each carried his own personal kit. The trail often led along narrow paths cut in the cliff face or on bridges that swung high above the tumbling torrent. These bridges were made of three ropes of twisted canes, one to walk on and two for hand-rails. It was difficult enough for the lightly laden climbing party to cross these swinging arcs, but for the porters with their heavy loads it was indeed a hazard. At noon the climbers passed the porters having their midday meal, and all reached camp before nightfall, one day's march nearer their goal.

17. Where were they going to?
   (on holiday : the Lake District : Mount Everest : camping)

18. What did each climber carry?
   (50 kg. : a haversack : a great load : his own kit)

19. What were the bridges made of?
   (stone : steel : twisted canes : concrete)

20. Who found it most difficult to cross the bridges?
   (the climbers : the leaders : the camels : the porters)

21. What time did the climbers pass the porters?
   (noon : dawn : 10 A.M. : in the evening)

22. The bridges were difficult to cross because . . .
   (they were dangerous : they were high up : they swung : they fell down)

23. How did they pass across the cliff faces?
   (by a bridge : at the top : on a man's back : on narrow paths)

24. How did the party travel?
   (in a train : in single file : in a bunch : on horseback)

In front of the combined harvester is a wide beater that turns slowly round, pressing the tall wheat-stalks firmly against the sharp, moving knives. As they fall the swathes of corn are collected by a moving belt and carried into the interior of the machine. Here the grain is beaten from the husk and then the smaller seeds and the chaff fall through shaking sieves.

The golden grain pours out in a steady stream into sacks which are dropped into a waiting lorry to be taken away to the granary, while the small seeds are collected in sacks at another chute. The straw is dropped in a cascade from the rear of the combine, while the chaff blows away in the wind as it pours out.

25. What are they harvesting?
 (barley : oats : wheat : maize)

26. What blows away in the wind?
 (the grain : the chaff : the straw : the small seeds)

27. What is collected in one set of sacks?
 (the straw : the small seeds the chaff : the stalks)

28. What does the moving belt do?
 (collects the grain : sacks the wheat : cuts the corn carries the cut corn into the machine)

29. What holds the corn against the knives?
 (the belt : the beater . the men : the sack)

30. What sorts out the small seeds?
 (the sack : the knives : the sieves . the belt)

31. What carries the sacks of grain away?
 (the belt : a cart : a man : a lorry)

English Test 2 *(15 minutes)*

Select from the brackets the word or phrase that means the same as, or almost the same as, the word or abbreviation at the beginning of the line.

1. *hurt* (injure : ignore : endure : allow)
2. *allow* (annoy : permit : employ : direct)
3. *e.g.* (any : for example : note well : see below)
4. *i.e.* (it was : it says : that is : that is all)

5. *nourish* (feel : fear : let alone : feed)

6. *stubborn* (obtain : deploy : prevent : obstinate)

Choose one word from each group in brackets to make the sentence correct.

7. When John comes (hear : here : heir : hare) you will see him.

8. What did you (sent : seen : send : sand) him for Christmas?

9. He plays the piano very (good : better : well : best)

10. Twenty men (was : were : wear : ware) killed.

11. (Both : Neither : Any) Tom (all : or : nor) I was there.

Select one word from those given in brackets that means the same, or almost the same, as the word at the beginning of the line.

12. *softly* (quietly : loudly : heavily : easily)

13. *finish* (languish : complete : forsake : deny)

14. *lift* (lower : lend : listen : raise)

15. *brave* (send : travel : conquer : valiant)

16. *vanish* (varnish : disappoint : disappear : deceive)

17. *control* (central : regulate : regale : sentry)

18. *foretell* (prophesy : prohibit : present : preclude)

What do we call the person who does each of these things?

19. Sells meat.

20. Mends water-pipes.

21. Flies an aeroplane.

22. Sells the tickets on a bus.

23. Sells cigarettes.

Alter the word in italic so that it completes the sentence when it is put into the blank space.

24. *cut* John was ____ the stick with his knife.

25. *black* It was the ____ night I have ever seen.

26. *carry* Tom ____ him on his back.

27. *make* They ____ him do it.

28. *day* He bought his ____ paper at the station.

29. *ask* He was always ____ questions.

30. *busy* He was in ____ as a grocer.

31. *depend* Tom is very ____ on his work.

What letters are needed to complete the words in boxes?

32. We went to the [Z | O | O | L | _ | G | _ | C | A | L] Gardens.

33. Do not [D | E | C | _ | _ | V | E] yourself.

34. The enemy were [D | E | F | _ | _ | T | E | D].

35. The lion hunts his [P | R | _ | Y].

36. The wolf returns to his [L | _ | _ | R].

37. He would not [R | E | L | _ | Q | _ | _ | S | H] his post.

Choose two words from each group in brackets that rhyme with the first word in the line.

38. *feet* (fleet : complaint : compete : relate : fate)

39. *bamboo* (refuse : blow : flew : confuse : flue)

40. *rail* (sell : retail : rental : near : tale)

41. *won* (one : tone : mention : win : fun)

42. *flee* (fly : see : flea : comply : read)

Choose the phrase that best explains the phrase in italic letters.

43. *to fit like a glove*
very apt
far too tight
most uncomfortable

44. *to be straight as a die*
reckless and foolish
thoroughly reliable
quite irresponsible

45. *a striking likeness*
a good clear hit
a glowing image
a close similarity

Choose the word from the group in brackets that is closest in meaning to the word in italic letters in the sentence.

46. It is difficult to say what his *motive* was.
(notion : reason : mention : anxious)

47. His manner was *arrogant* towards his inferiors.
(frightening : cunning : haughty : avaricious)

Copy out these sentences, putting in punctuation marks and capital letters.

48. tom asked can i go now

49. the *guardian* is a daily paper

50. can you tell me said mary if he will come

English Test 3 *(15 minutes)*

Down the dark well swung the bucket with Bob standing in it and clinging to the rope with all his might. As soon as he flashed the signal Tom followed, then came Ned and Tim in quick succession. When they were all down Tim led them along the twisting passage that led into the first cave. Here the water of the underground lake lapped their feet as they paddled along the sunken ledge that led to the treasure cave. The chest of gold coins lay in the far corner and it took them some time to locate it.

1. Who led the way down the well?
   (Tim : Tom : Bob : Ned)

2. What was there in the first cave?
   (the treasure : a lake : a well : a bucket)

3. How did they reach the second cave?
   (through a passage : down a well : along a sunken ledge : round a corner)

4. Who came down the well last?
   (Tim : Tom : Ned : Bob)

5. How did Bob signal to the others?
   (he shouted : he pulled the rope : he flashed a light : he called up)

6. What did Bob hold on to as he came down?
   (a rope : a bucket : a well : a hook)

7. How many boys went down the well?
   (one : two : three : four)

8. What were they looking for?
   (a man : a well : a chest : a light)

And then in a moment it happened! It was not the chain that broke, but it was the collar that slipped, for it had been made for a thick-necked Newfoundland. We heard the rattle of falling metal, and the next instant dog and man were rolling on the ground together, the one roaring in rage, the other screaming in a strange shrill falsetto of terror. It was a very narrow thing for the Professor's life. The savage creature had him fairly by the throat, its fangs had bitten deep, and he was senseless before we could reach them and drag the two apart. It might have been a dangerous task for us, but Bennett's voice and presence brought the great wolfhound instantly to reason.

9. How did the dog get loose?
   (the chain broke : he slipped out of his collar : they let him go : he came from behind)

10. Who screamed?
(the Newfoundland : Bennett : the Professor : the dog)

11. What made it easier to make the dog let go?
(he was tired : he did not hold very tightly : his fangs bit deep : Bennett calmed him down)

12. What sort of a dog was it?
(a wolfhound : a Newfoundland : a terrier : a spaniel)

13. How did we know the dog was off his chain?
(we saw him slip his collar : he broke his chain : we heard the chain fall : we saw his collar was undone)

14. Whom did the dog attack?
(Bennett : the Newfoundland : a small dog : the Professor)

One bright person, in those far-off days when our streets were full of horses, came forward with a simple idea for stopping a runaway. On the cart fixed to this animal you put a lifting-apparatus worked by a wheel turned by hand. This lifting-apparatus has a hook which is fixed to a leather band at about the middle point of the horse. If the horse gets ideas into his head, and comes to a decision to get away from the place as quickly as possible, you simply give the hand-part of the wheel a turn. The animal is then lifted up by the apparatus till his feet are no longer on the earth.

15. Which title best describes this paragraph?
(the cart before the horse : why cars were invented : a stupid invention : a horse with ideas)

16. What did you have to do to lift the horse?
(raise your hand : fix a leather band round the horse : turn a wheel : stop him)

17. What was the hook fastened to?
(a leather band : a cart : a wheel : a lifting-apparatus)

18. What does the horse getting ideas in his head mean?
(he thought a lot : he is lifted off the ground : he decides to run away : his feet are off the ground)

Woody nightshade grows plentifully in hedgerows. It is a tall scrambling plant which usually supports itself on stronger plants. Its small purple flowers grow in loose clusters, and there are two green spots on each petal. Deadly nightshade is quite a different plant, very poisonous and rather rare. You will know it by its large, solitary, drooping, bell-shaped purple flowers and its big black poisonous fruits which are as large as cherries. The fruits of the woody nightshade are scarlet in colour, much smaller in size, and hang in clusters from the stem. The deadly nightshade plant stands strongly by itself. There are two other sorts of nightshade.

19. How does woody nightshade support itself?
(on trees : on ledges : on stronger plants : on walls)

20. What hangs in clusters from woody nightshade?
(leaves : its red flowers : its petals : its red fruit)

21. Why is deadly nightshade dangerous?
(it is rare : its leaves tear your clothes : its flowers are red : its fruits are poisonous)

22. Which has the smaller flowers?
(woody nightshade : deadly nightshade)

23. The flowers of the woody nightshade are . . .
(small and red : large and purple : red and spotted : purple and spotted)

24. The fruit of the deadly nightshade is . . .
(black and large : red and large : small and yellow : large and brown)

25. Which flowers hang like bells?
(woody nightshade : deadly nightshade)

If you found four plants answering these descriptions, which would be woody nightshade?

(purple solitary flowers and red berries)
(large black berries, climbing up a hedge)
(purple flowers in clusters and small berries)
(purple bell-shaped flowers with two green spots on each petal)

English Test 4 *(25 minutes)*

Name the people who do these things:

1. Carry luggage at the station.
2. Look after sheep.
3. Ride horses in races.
4. Repair shoes.
5. Guide ships into harbour.

Choose a word from each group in brackets to complete these phrases satisfactorily.

6. as white as (a sheet : a tree : a vase : a pine)
7. as bold as (milk : gold : brass : a baby)
8. as clear as (crystal : sand : a pond : a forest)
9. as clever as (a donkey : a pig : a sheep : a monkey)
10. as dry as (water : dull : dust : bread)

Choose the correct word from each pair in brackets.

11. They had (there : their) uncle with them.
12. What will you (wear : were) to go to school?
13. Tom is the (biggest : bigger) of the two.
14. Fred and (I : me) (was : were) there.

Choose one phrase that means the same as the one printed in italic in the sentences below.

15. He showed the *white feather.*
    he kept chickens
    he was a coward
    he played Red Indians

16. They decided to let *sleeping dogs lie.*
    to keep quiet about it
    to stop teasing the dog
    to leave it to the dog

17. The boys decided to *bury the hatchet.*
    to be good
    to have a fight
    to make peace

Choose the word from the group in brackets that means the same as, or nearly the same as, the word at the beginning of the line.

18. *courtesy* (politeness : courtly : sensation : slovenly)

19. *peculiar* (precious : pertinent : poor : strange)

20. *exhibit* (assess : explode : conclude : display)

21. *dispute* (delight : argument : arrange : descry)

22. *imagine* (think : plan : fancy : complete)

23. *abundant* (plentiful : absurd : frequently : about)

Choose the correct spelling, from the four spellings given, of the word between brackets.

24. It is (necessery : necessary : nesessary : necesary) for you to know how to spell.

25. I am only sending you a (breif : brief : breef : breaf) note.

26. Tom (persauded : persuded : persuaded : pursuaded) Mary to come for a walk.

Make a word from the word at the beginning of the line to fill in the blank in the sentence.

27. *winter* It was a cold and __________ day.

28. *survive* There was only one __________.

29. *reduce* There was a ________ on every article in the sale.

30. *assist* My ________ was not needed.

31. *object* The top of the mountain was their ________.

Rewrite these sentences in the **plural.**

32. I want to go to my home.

33. One book fell from the table.

34. The boy was climbing over the wall.

35. The horse jumps the fence.

Choose one word from each group in brackets to replace the phrases in italic in these sentences.

36. The King *gave up his throne.*
    (abdicated : deposed : traitor : coward)

37. Mary and I went to *the place where horses are shod.*
    (barn : smithy : garage : stable)

38. He made this mistake through *knowing no better.*
    (foolishness : injustice : ignorance : ignore)

39. We crossed the river by the *small boat that carries people across the river.*
    (canoe : ferry : steamer : raft)

40. The glass was *the sort you cannot see through.*
    (clear : opaque : transparent : obtuse)

41. The *man who sells paints, screws, and paraffin* sold me some nails.
    (grocer : tinsmith : ironmonger : tinker)

42. The *man who tells the pilot which way to steer* told Jim to go due north.
    (plotter : navigator : co-pilot : stewardess)

43. He was found guilty of *telling lies under oath.*
    (prejudice : perjury : lying : swearing)

Complete each sentence with an apt word replacing each dot by a letter.

44. Tom is very s . . . . m out late at night.

45. Mary is c . . . . . s and inquisitive about things.

46. Their leader e . . . . d them to work better.

47. He made a very good im . . . . . ion on the committee.

Copy these sentences, putting in capitals and punctuation marks.

48. he shouted help help i am drowning

49. halt cried the sentry who goes there

50. mary and fred with all speed flung up the window

English Test 5 *(15 minutes)*

Tom and Mary hid in the beech-tree until tea-time. By then they were very hungry but they were still afraid to come down. However, at last hunger got the better of fear and they dropped quietly to the ground. The wood seemed very still as they made their way slowly from tree to tree expecting at any moment to come face to face with the great ape. However, they reached the open fields without mishap and raced home to Fell Farm as fast as their legs would carry them.

1. What had frightened Mary and Tom?
   (a mouse : an ape : a bull : a farmer)

2. Where did they live?
   (in a tree : in a wood : at a farm : in a town)

3. What sort of a tree did they hide in?
   (elm : ash : oak : beech)

4. How did they go through the wood?
   (they ran : they crept : they walked : they raced)

5. Where did they come to when they left the wood?
(home : the farm : the trees : open fields)

6. What time did they leave the tree?
(noon : tea-time : bedtime : in the morning)

7. What made them leave the tree?
(the ape : fear : hunger : their father)

8. Why did they not come down from the tree before?
(they were afraid : it was tea-time : it was late : they did not want to)

It was a wild morning in October, and I observed as I was dressing how the last remaining leaves were being whirled from the solitary plane-tree which graces the yard behind our house. I descended to breakfast prepared to find my companion in depressed spirits, for, like all great artists, he was easily impressed by his surroundings. On the contrary, I found that he had nearly finished his meal, and that his mood was particularly bright and joyous, with that somewhat sinister cheerfulness which was characteristic of his lighter moments.

9. What time of year was it?
(spring : summer : autumn : winter)

10. What sort of a tree was there behind the house?
(an oak : a beech : an ash : a plane)

11. What was my friend influenced by?
(his breakfast : the plane-tree : his lighter moments : his surroundings)

12. Why was I surprised?
(the leaves were falling : my friend had not started his breakfast : my companion was cheerful : there was nobody there)

13. What did I notice as I dressed?
(my companion was depressed : the last leaves were falling : that my friend had nearly finished his breakfast : it was October)

The wall was built by the Romans almost two thousand years ago to keep the wild Picts from raiding the settled lands of the south. It stretched for one hundred and thirty kilometres across the country from the North Sea in the east to Solway Firth in the west. It was made from trimmed stone and was two metres high. On the north it had a parapet one metre high. Every two kilometres there was a castle for the guards and signal towers were spaced at intervals between these. Garrison towns were built behind the wall to the south where the soldiers rested when they were not on duty.

14. Who built the wall?
    (Romans : Picts : British : enemies)

15. Where did the soldiers live when they were resting?
    (in castles : in garrison towns : in signal towers : on the wall)

16. How high was the parapet?
    (3 m. : 1 m. : 1.5 m. : 2 m.)

17. Which side of the wall were the garrison towns?
    (north : south : east : west)

18. Why was the wall built?
    (to keep out the Romans : to keep out the soldiers : to keep out the settlers : to keep out the Picts)

19. What sort of stones was the wall built of?
    (hard : rough : trimmed : soft)

20. What lay at the west end of the wall?
    (the enemy : the Solway Firth : the Picts : the North Sea)

21. What were the towers for?
    (Guard-houses : to rest in : to see from : to signal from)

The moon revolves round the earth, and both of these revolve round the sun, as do the other planets such as Saturn, Mars, and Venus. In spite of this constant rotation it is seldom that any three of these heavenly bodies come into line with each other. When they do there is an eclipse.

Since the other planets are so far from us we notice their presence in an eclipse very little, but the moon is much nearer. It can either totally eclipse the sun for us by passing between the earth and the sun (an eclipse of the sun), or be itself eclipsed by the earth when the earth comes between it and the sun (an eclipse of the moon). We seldom see eclipses of the sun but eclipses of the moon are more frequently observed.

22. What does Mars revolve round?
    (Venus : Saturn : Sun : Earth)

23. Which eclipses are more often seen?
    (Moon : Earth : Venus : Mars)

24. There is an eclipse when . . .
    (the moon meets the earth : three heavenly bodies are in line : the moon goes behind the sun)

25. We call it an eclipse of the sun when . . .
    (the sun goes behind the moon : the earth covers the sun : the sun sets : the moon goes behind the earth)

26. Why do we not notice an eclipse of Mars?
    (it never happens : it is always happening : Mars is so far away : you cannot see it)

27. An eclipse of the moon is caused by . . .
    (the earth passing between the sun and the moon : the sun passing between the earth and the moon : the moon passing between the sun and the earth)

English Test 6 *(30 minutes)*

Which of the four words below each sentence means the most nearly the same as the word in italic letters in the sentence? Write down the letter which is in front of the word you choose.

1. The flag *waved* at the mast-head.
   (*a*) fluttered (*b*) drooped (*c*) blue (*d*) was fastened

2. Tom ran *quickly* down the road.
   (*a*) slowly (*b*) ambled (*c*) swiftly (*d*) quietly

3. He is an *awkward* fellow.
   (*a*) awesome (*b*) stupid (*c*) clumsy (*d*) strange

4. He *often* went for a walk at night.
   (*a*) seldom (*b*) frequently (*c*) sometimes (*d*) never

5. The *apprentice* studied his craft.
   (*a*) learner (*b*) youth (*c*) artisan (*d*) worker

Pick out the word that could be used in place of the given phrase and write down the letter which is before it.

6. a man who sells bread and cakes
   (*a*) ironmonger (*b*) baker (*c*) butcher (*d*) stationer

7. a room where books are kept
   (*a*) hall (*b*) kitchen (*c*) library (*d*) book-worm

8. a piece of ground where small quantities of many kinds of vegetables are grown
   (*a*) park (*b*) field (*c*) farm (*d*) allotment

9. not guilty
   (*a*) convicted (*b*) gaoled (*c*) let off (*d*) innocent

10. to get faster and faster
   (*a*) speed (*b*) rallentando (*c*) accelerate (*d*) accentuate

Choose the word or phrase in brackets that means the same as the phrase above it.

11. hit below the belt
   (take unfair advantage : turn the other cheek : eat too much)

12. I wash my hands of it
   (I forget what I meant : I do not do my job : will have no more to do with it)

13. hit the nail on the head
   (shout before you are hurt : shut the stable door after the horse is stolen : be absolutely right)

14. beat about the bush
    (go hunting : hit the wrong man : talk round the matter)

15. burn the candle at both ends
    (waste light : burn your fingers : try to do too much).

16. put your foot in it
    (lose you dignity : make a mistake : step off with the wrong foot)

17. sit on the fence
    (be unable to make up your mind : be lazy : be two-faced)

Choose the word in brackets that best fits each sentence

18. The enemy crept (loudly : stiffly : cautiously : lazily) across the field.

19. After the trees had been cut down the land was (forested : reafforested : forest-clad : forestry).

20. Tom (came : enlisted : enlightened : settled) in the Navy.

21. Her fine clothes made her seem very (warm : casual : component : pretentious).

22. The hillside was (felled : covered : denuded : demolished) of timber.

Take one phrase from each of the two columns given below which make a sentence that means the same as the two sentences given. Write down the letters that show the phrases you choose.

23. The snake poisoned the man. He died of poison.

(*a*) The man who was bitten by the snake	(*e*) so he died of poison
(*b*) The snake who bit the man	(*f*) that it killed him
(*c*) The poisoned man	(*g*) died of poison
(*d*) When the snake bit the man	(*h*) that he died of poison

24. I ran into the field. The gate was open.
   (*a*) The field I ran into
   (*b*) I ran into a field
   (*c*) I ran into the field
   (*d*) I ran from the field
   (*e*) the gate was open
   (*f*) had the gate open
   (*g*) the gate was opened for me
   (*h*) from which the gate was open

25. The boy ran fast. The race was won by him.
   (*a*) The fast running boy
   (*b*) So fast the boy
   (*c*) The boy ran so fast
   (*d*) The fast boy
   (*e*) that he won the race
   (*f*) so he won the race
   (*g*) because he won the race
   (*h*) and he won the race

Rewrite these sentences starting with the words followed by dots.

26. That he did it is my opinion.
   In my opinion . . .

27. Gardening is enjoyed by most men.
   The majority . . .

28. Cycling is quicker than running.
   A man can . . .

29. He fell because he ran too fast.
   The cause of . . .

30. There is no doubt that butter is better than margarine for cooking.
   The superiority . . .

Write these sentences in the past tense.

31. I want you to write carefully.

32. Where will you go now?

33. I leave you now to see what you will do.

34. I will come as soon as I can.

35. Many great animals are crossing the river.

Choose one word or phrase from those in brackets that means the same as the word at the beginning of the line.

36. *crude* (unknown : lacking finish : credulous : not understood)

37. *prohibited* (too awkward : too expensive : too loud : not allowed)

38. *complacent* (self-satisfied : unlucky : full up : full of congratulations)

39. *belligerent* (a party of peace : ringing the bells : waging war : a drooping bell-like flower)

40. *candid* (sweet and sugary : always there when wanted : outspoken : persuasive)

Use a word that is formed from the word at the beginning of each line to fill the gap in each sentence that follows.

41. *inform* He was spied upon by an ________.

42. *complete* On the ________ of the house I shall live in it.

43. *round* He is used to ________ off his sentences.

44. *force* He had a very ________ character.

45. *contempt* Tom seemed to be ________ of his brother.

Copy these sentences, putting in capitals and punctuation marks.

46. where do you come from enquired lily

47. is it a treacle well asked alice

48. i feel most unwell said the king

49. come on shouted the crowd shoot

50. of course i will i said if you will let me

Composition

Some typical subjects:

1. My home.
2. My most exciting adventure.
3. The room I am sitting in.
4. How to make a fire.
5. "I looked up and saw him climbing out of the window . . ."
   Make up a story starting with these words.
6. The best story I know.
7. Write a story which finishes:
   ". . . and I woke to find myself on the floor."
8. My favourite book.
9. "How I ran away" by a bicycle.
10. "My first win" by a pony.
11. When the chimney caught fire.
12. Write a story that ends with a sentence which begins:
    "If I had come five seconds later . . ."
13. Said the pudding to the spoon.
14. This is the finest deed I remember.

Allow 30–40 mins. for the child to plan and write the composition. When you read it look for the following points.

1. Has it a good beginning and a good end?
2. Does it keep to the point?
3. Is it interesting?
4. Are the words used apt and well chosen?
5. Look for serious grammatical errors.
6. Look for spelling mistakes.
7. Check punctuation.

Arithmetic Test 1 *(25 minutes)*

Add (+)

1.
```
 393
 487
 129
```

Subtract (−)

2.
```
5643
2952
```

Multiply (×)

3.
```
476
  7
```

Divide (÷)

4. 4297 by 7

Add (+)

5.

m.	cm.	mm.
1	90	7
3	1	8
2	10	0

Multiply (×)

6. 907·6 g. by 8. (Answer in kg.)

Subtract (−)

7.
```
 £
5·78
2·59
```

Divide (÷)

8. t 53 kg 15 by 5 (Answer in tonnes)

Multiply (×)

9.
```
 £
5·25
   5
```

Add (+)

10. 7·49 l., 3·24 l., 0·76 l. and 720 ml.

Divide (÷)

11. hr. 9 min. 35 sec. 25 by 6

Subtract (−)

12.
```
  £
12·65
 6·75
```

Multiply (×)

13.

hr.	min.	sec.
2	15	10
		8

Add (+)

14.
```
  £
 4·25
 0·90
17·60
```

Divide (÷)

15. 27·441 kg. by 9.

Add (+)

16.

hr.	min.	sec.
3	25	18
2	47	3
1	58	35

Multiply (×)

17. 7·2016 l. by 7.

18. Subtract (−)
£
1·32
0·50½

19. Divide (÷)
t. kg.
15 215 by 10
(Answer in kg.)

20. Divide (÷)
1·33 l. by 7
(Answer in ml.)

21. Multiply (×)
kg. g.
11 195
9
(Answer in kg.)

22. Subtract (−)
1000
296

23. Divide (÷)
£
18·45 by 9

24. Subtract (−)
week day hr.
5 3 7
2 6 10

25. Divide (÷)
hr. min. sec.
49 15 55 by 8

26. Divide (÷)
16·045 t. by 5
(Answer in kg.)

27. Multiply (×)
£
5·75
7

28. Multiply (×)
0·045 m. by 9
(Answer in cm.)

29. Subtract (−)
km m
100 0
57 708
(Answer in km.)

30. Divide (÷)
0·96 l. by 5 (Answer in ml.)

Arithmetic Test 2 *(25 minutes)*

1. What must I multiply 7 by to get 35?
2. What would 5 metres of ribbon cost at 5p for 50 cm.
3. If a page has 15 lines and each line has eight words on it, how many words will there be on a page?
4. If 16 boys collect 48p between them and they share it out equally, how much does each receive?
5. How many kilograms are there in a quarter of a tonne?
6. What is an eighth of 2 kg. in grams?
7. Which of these numbers is one-sixth of 96?
   12, 16, 18, 22, 24.
8. What is the average of 8, 10, and 12?
9. I am 40 cm. shorter than Fred and he is 1·6 m. How tall am I?
10. How much heavier than 175 g. is ¼ kg.?

11. How much is 250 g. of tea at £2.50 a kg. and 500 g. of coffee at £5.40 a kg.?
12. If I spent 74p, how much change should I have from £1?
13. If I have 500 g. of apples and eat 400 g. of them, what fraction of 500 g. have I left?
14. If a leg of lamb weighs 2 kg. and costs £3·80, how much a kg. is it?
15. Write in figures three hundred and twelve.
16. If Mary is twice as old as Tom by all but three years and she is seventeen, how old is Tom?
17. What is the difference between a score and a dozen?
18. How many pieces of wood, each 35 cm. long, can I cut from a plank 3 m. long?
19. If Mary gives 6 marbles to Betty and Betty gives 7 marbles to Jane and Jane gives 4 marbles to Mary, who then has 20 marbles, how many marbles had Mary to begin with?
20. What number divided by 7 gives 8?
21. At noon on October 17th, how many days is it to noon on Christmas Day?
22. What fraction of 1 kg. is 450 g.
23. If I sell a dozen hens for £26 and lose £1·60, how much did I pay for each hen?
24. If there are ten rows of trees in an orchard and ten trees in each row, how many trees are there in the orchard?
25. If a trawler lands 40 baskets of fish each weighing 50 kg. , how many tonnes of fish does it land?
26. An aeroplane travels at 1,200 km./h. How far will it go in 20 min.?
27. How many egg-boxes each holding a dozen do you need to pack 204 eggs?
28. How many ⅓ l. bottles of water would be needed to fill a ten-litre tank?
29. On an average a flock of 80 sheep has 0·75 of a lamb each. How many lambs did the flock have?
30. A panel of woven wood fencing is 1 m. long. How many panels do I need to fence my rectangular garden? The longer sides are 18 m. long and the shorter sides 10 m.?
31. A bicycle costs £30 down and £1·50 a month for 2 years. What do I have to pay for it in all?
32. 144 boys went to play soccer. If each team must have a reserve, how many teams can they make?

33. A boy cycles at 25 km./h. How long will it take him to cover 65 km.?
34. How much can I save by buying 250 logs at £18.00 for 50 instead of 50p each?
35. On 17 hectares a farmer grows an average of 5 tonnes of wheat on each hectare. How much wheat has he to sell?
36. What is ½ of ¾?
37. If eggs are 84p a dozen and a hen lays 6 eggs a week, how much will the eggs she lays between March 5th and April 8th be worth?
38. How many times must a boy run the full length of a cricket pitch to cover a mile?
39. In a fishing competition 56 men catch 48 kg. of fish. On an average how much is this each?
40. A boy covers 16 km. by car, bus, and walking to reach school. The car carries him 3½ km. and the bus travels for 20 minutes at 30 km./h. How many kilometres has he left to walk?
41. A man receives one and a half times his hourly rate of pay for working over Christmas. If he is normally paid at £7.20 an hour and works for fifteen and a half hours, how much does he earn over Christmas?
42. If I join end to end three pieces of wood 8 cm., 5·75 cm., and 2·25 cm. long and then cut off a quarter of the whole length, how much is left?
43. If the attendance (morning and afternoon) for a class of 40 children for a week is:
    38, 39, 36, 34, 38, 39, 37, 36, 36, 37.
    What is the average attendance for the week?
44. In a row of holiday huts there are two people in each even-numbered hut and three people in each with an odd number. How many will there be in huts numbered 1 to 60?
45. A car radiator holds 18 litres of water. How many litres are left when half the water has boiled away?

Mathematics Test 1 *(30 minutes)*

Arrange these figures so that they make:

1. the biggest possible number;
2. the smallest possible number:
   2 8 3 7 8.
3. What is the first number in this series?
   __ __ __ __ 22, 24, 26, 28.

4. Sixty thousand five hundred and twenty-seven people paid to watch a football match. If admission tickets were in books of a hundred, how many books were needed?
5. A shelf of 27 books was shared among 9 boys for the holidays. How many did each boy have?
6. A dozen bulbs cost 45p. Only $3/4$ of them bloomed. How much did each bloom cost?

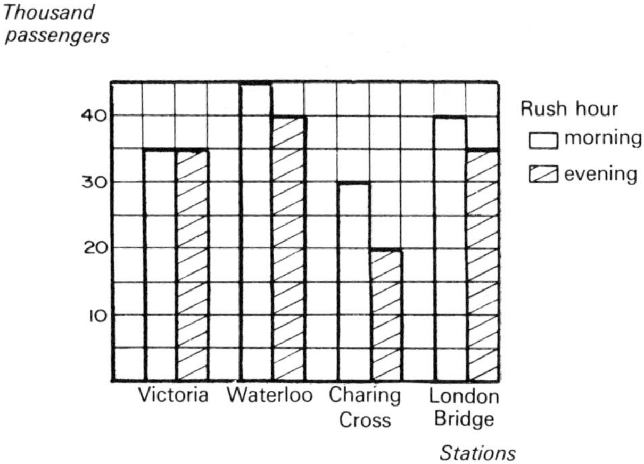

7. Which station is busiest in the morning rush hour?
8. How many more passengers use Victoria than Charing Cross in the evening rush hour?
9. How many passengers use all these stations during the morning rush hour?
10. If a train carries 300 passengers on an average, how many trains are needed for the morning rush hour?
11. How many trains are needed for both the rush hours at Waterloo?
12. What is 25% of 9.72 kg.?
13. What decimal fraction of 1 kg. is 500 g.?
14. What fraction of a metre is 250 cm. (in lowest terms)?

There are 100 small squares in this pattern.

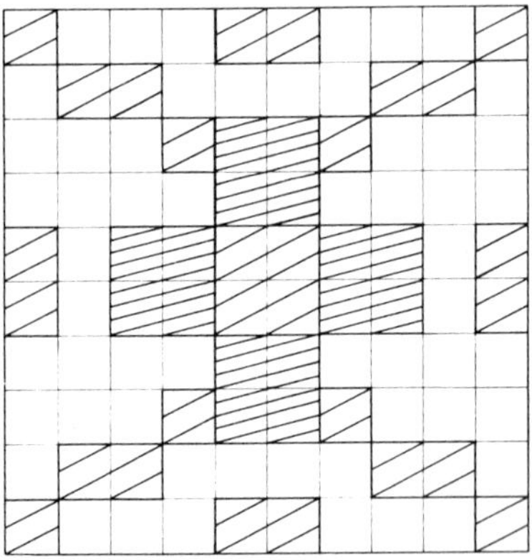

15. What percentage of them is darkly shaded?
16. What decimal fraction of them is unshaded?
17. What fraction of them is shaded?
18. How many more of them would have to be shaded to leave only a half unshaded?

Fill in the blanks:

19. $24 \div 12 = 48 \div$.
20. $32 \times 24 = \quad \times 8$.

Add two to each of these series:

21. 4, 8, 12, 16, , .
22. 1, 2, 4, 7, , .
23. Dick can cycle 110 m. while Tom cycles 120 m. How far behind Tom will Dick be when Tom has covered 1 km. and 800 m.?
24. Sweets cost 1p, 2p, and 4p each. If I buy eighteen sweets, an equal number of each sort, how much do I have to pay?
25. Arrange these fractions in order, starting with the smallest:

$\frac{4}{7} \quad \frac{3}{5} \quad \frac{1}{12} \quad \frac{1}{2} \quad \frac{9}{10}$

This door is made of four lengths of wood 25 cm. wide:

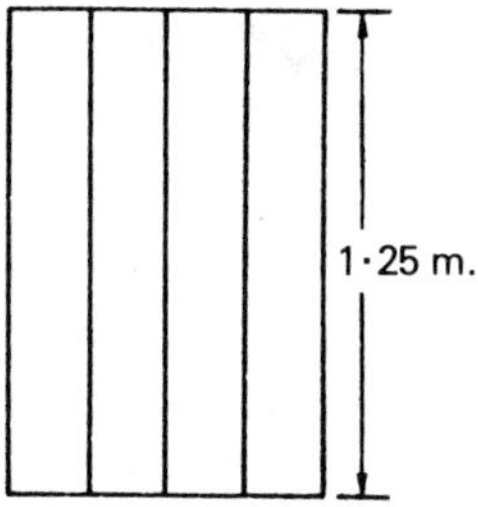

26. What length of 25 cm wood is required?

27. What is the area of the door?

28. If a 10 cm. strip were added all round it, what length of wood would be needed?

29. In a school there are six classes with 29, 31, 33, 35, and 27 children and one class whose number has been left out. We know that the average for the six classes is 30. What is the missing number?

30. If a round cake is cut so that Joan has half and Peggy has a quarter, what is the angle at the centre of the cake of the piece that is left?

Mathematics Test 2 *(30 minutes)*

1. Multiply 3·45 m. by 8.

2. Subtract 176 g. from 312 g.

3. Divide 5·6 l. by 4 and write your answer in ml.

Fill in the blanks:

4. $5+4=3\times$.

5. $2+\quad=16\div4$.

6. $6+2=4\quad2$.

7. $8\quad1=32\div4$.

8. If $345\div15=23$, then $23\times15=$.

9. At 60 km./h. how far shall I travel in 20 min.?

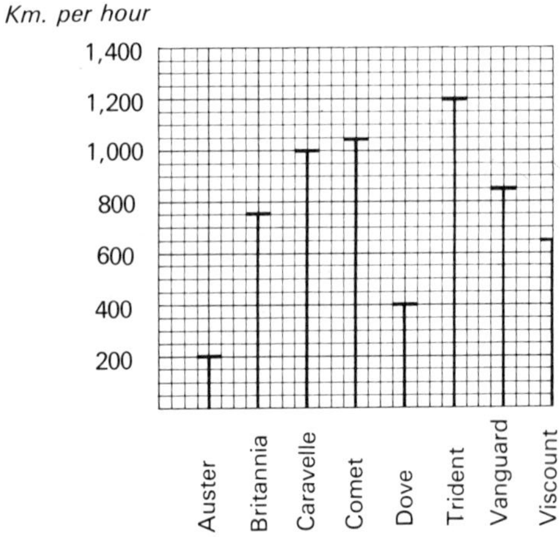

10. Which is the fastest plane?
11. How much faster is the Vanguard than the Viscount?
12. If I travel an hour by Auster, then three hours by Comet, and then an hour by Britannia, how far have I gone?
13. When two numbers are multiplied together, the answer is one thousand. If one number is 10, what is the other number?
14. How many complete tens are there in 52,684?

Fill in the blanks:

15. ½ = /12.
16. 4¾ = /4.
17. A sewing-machine costs £90 if you pay cash. On instalments you pay £12 down and £7.25 a month for a year. How much do you save by paying cash?
18. A dining-table cost £127, six chairs cost £22·50 each, and a sideboard costs £135. Which of the figures given below is likely to be nearest to the total cost?
£500, £200, £350, £450, £400.

This shows a shopkeeper's takings day by day during the week:

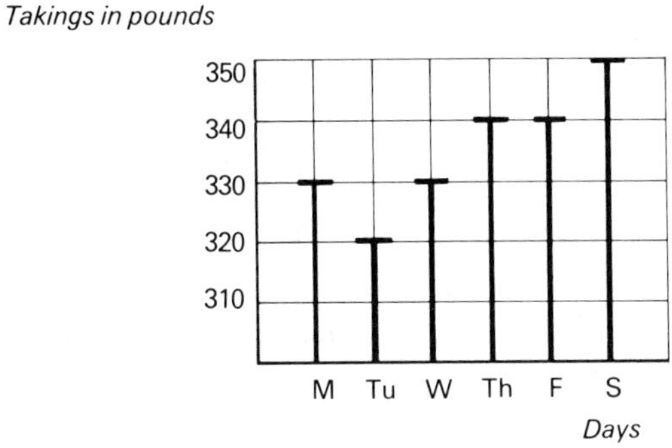

19. Which day did he take least?

20. How much did he take for the whole week?

21. What were his average daily takings?

22. $1 \cdot 5 = 1\frac{1}{2}$

Pair up these:
$1 \cdot 9$, $1\frac{3}{4}$, $1 \cdot 2$, $1\frac{9}{10}$, $1\frac{1}{5}$, $1 \cdot 75$.

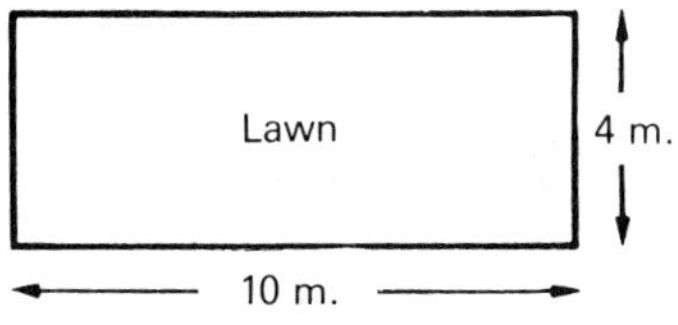

23. How many slabs each $0 \cdot 5$ m.^{2} do I need to make a path all round this lawn?

24. What is the difference between the fours in this:
34, 564.

25. When two numbers are multiplied they give 32. When the larger is divided by the smaller the answer is 2. What is the smaller?

This graph shows that a man walked for four hours at 6 km. an hour:

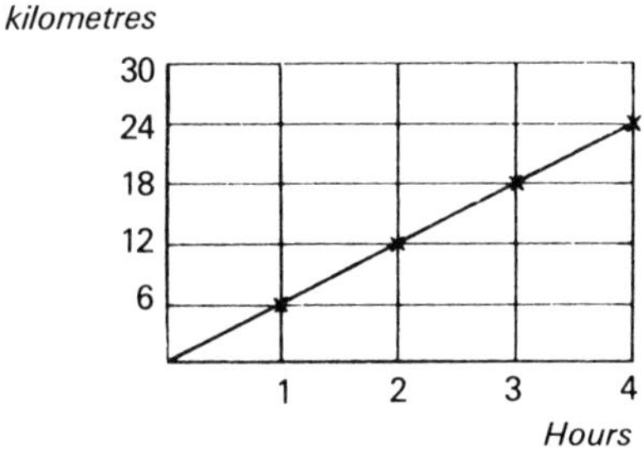

26. How far did he walk in 4 hours?
27. How far in ½ hour?
28. How far in 3½ hours?
29. 1/6 of the boys in the class have red hair. If there are 5 boys with red hair, how many boys are there in the class?

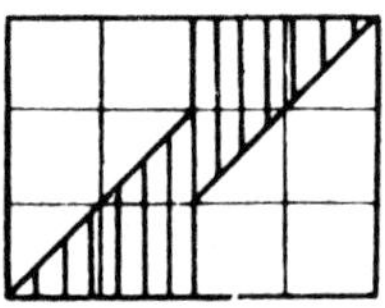

30. On the same pattern of small squares draw a square equal to the two triangles shaded here.

Mathematics Test 3 *(30 minutes)*

	Multiply (×)		Add (+)		Subtract (−)
1.	12 kg. 142 g. by 7 and answer in kilograms.	2.	2·419 m., 3·828 m., 0·16 m. and 0·092 m.	3.	min. sec. 40 15 39 45

One of the signs has been left out of each of these sums. What is it?

4. 2 2=4.
5. 6 3=18.
6. 5 4−3=6.
7. 6 2−2=10.
8. ½ ½=¼.
9. 3 5+1−2=7.

10. What number divided by 6 gives 3?

11. A boy cycles at 30 km. for $2\frac{1}{2}$ hours. How far does he travel?

12. If there are 16 lights in a house and three-quarters of them are off, how many are alight?

13. If 1 tonne of coke keeps a furnace burning for 5 days, how many kg. are burned each day?

Without doing any working, pick out the largest of these:

14. $57-42$, $57+42$, 57×42, $57\div42$.

15. $1\cdot7$, 17, $^{1}/_{17}$, $0\cdot17$.

16. $0\cdot7777$, $^{177}/_{122}$, $^{177}/_{777}$, $^{177}/_{2}$.

17. If I buy a house for £45,000 and pay £7,500 for a garage and £2,250 for curtains and carpets, how much have I paid altogether?

18. How many 125 g. packets of chocolate are there in a box that holds 7 kg. of chocolate?

	Divide (÷)		Multiply (×)
19.	48,291 by 9	20.	4,164
			49

By looking at this multiplication sum answer the following questions:

$$\begin{array}{r} 453 \\ \underline{42} \\ 906 \\ \underline{18{,}12} \\ 19{,}026 \end{array}$$

21. $\frac{19{,}026}{42}$.

22. 453×4.

23. 453×21.

24. $19{,}026-18{,}120$.

In 1958 an American submarine, *Nautilus,* travelled under the Arctic Ocean and the North Pole on a journey from Hawaii to Washington, U.S.A.

Here are some facts about this remarkable voyage:

Left Hawaii July 23rd, 07·00.
Dived under ice August 1st, 13·37.
Reached North Pole August 4th, 04·15.
Surfaced from dive August 5th, 14·50.
Reached Washington August 8th, 15·15.

25. Turn 07·40 to a.m. or p.m.

26. Turn 14.50 to a.m. or p.m.
27. How many hours and minutes was the submarine under the ice?
28. How much longer was the journey under the ice to the North Pole than the journey back to the open sea again?
29. How long did the journey from Hawaii to Washington take in days, hours, and minutes?

30. Multiply (×)		31. Subtract (−)		32. Divide (÷)
15 t. 170 kg. 300 g. by 9 and answer in tonnes.	31.	15·209 l from 20·104 l.	32.	£ 59·76 by 8.

33. A lorry and its load of six crates weighs 6,700 kg. The lorry weighs 2,500 kg. How much does each crate weigh?
34. Two numbers added together make 10. One is 2 larger than the other What is the smaller?
35. Which is the smallest of these fractions?

    $\frac{2}{5}$, $\frac{1}{3}$, $\frac{1}{10}$, $\frac{3}{4}$, $\frac{5}{15}$.

36. Travelling by plane, you are allowed 20 kg of luggage free, but you must pay 30p per 500 g above this. How much would you pay on 34 kg.?

Mathematics Test 4 *(40 minutes)*

1. Draw the flag that is flying on both lines.

2. What is the difference between 3 and 9?
3. Write four and a quarter.
4. How many chocolates are there in seven boxes with two rows of three in each box?
5. What is the average of 96, 121, 78, 65.
6. How many triangles like this can be cut from the rectangle?

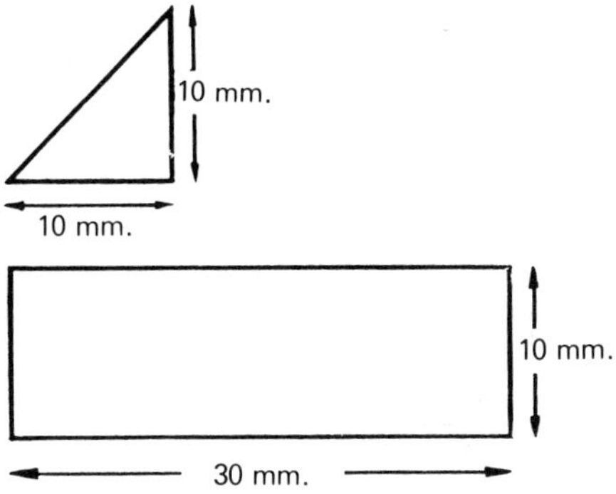

7. Write four units and four-tenths as a decimal.
8. How many times is the first three in this number bigger than the second?

   3143.
9. Write £3 and seventy-five pence in figures.
10. Add £2 and 10p, 45p and 6p.
11. How many boys keep both rabbits and mice?

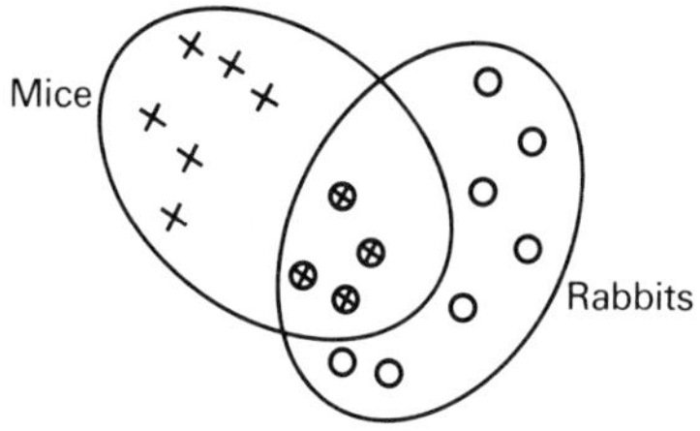

This shows how much space, measured in total length of columns, the local newspaper gives to different topics.

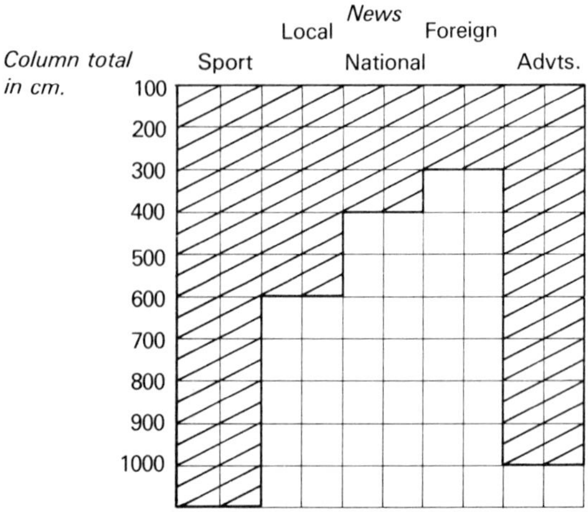

12. Which has the greatest length?
13. How much more local than national news is there?
14. How much of the paper is taken up with advertisements?
15. Draw a triangle and a circle that overlap. Write the numbers one to nine so that the three larger numbers fall in the circle, the three smaller in the triangle, and the others in both.
16. Pick out all the prime numbers in this line.
    7, 18, 23, 32, 33, 47, 50.

What is missing from these sums?

17. 3 2=5.
18. 33 3=11.
19. 5 5=25.
20. ½+ /6=1.
21. ¼ ¼=1.

What figures are missing from these sums?

22.
```
   47
+ 987
 1  3
```

23.
```
  678
+35 9
  827
```

24.
$$\begin{array}{r} 73 \\ -\ \ 8 \\ \hline 111 \\ \hline \end{array}$$

25.
$$\begin{array}{r} 731 \\ - \\ \hline 259 \\ \hline \end{array}$$

Write down the largest number in questions 26 to 28:

26. $\frac{1}{3}$, $\frac{1}{5}$, $\frac{1}{2}$, $\frac{1}{10}$.

27. 3, 3·3, ·33, ·3.

28. ·35, ·035, ·351.

29. 4·34×3·2.

30. 94·2÷3.

31. 1·044÷·36.

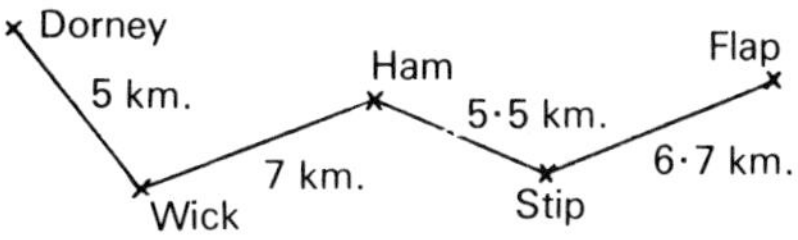

32. How far does a postman travel to deliver letters to these four villages?

33. What is the average distance between the villages, to the nearest metre?

34. If T and O were turned round their shape would be the same. With which of these letters can you do the same?
    M, S, L, N, A, C, X.

35. If a toy costs 27½p and you give a 50p piece, how much change would you expect?

36. If you are given the smallest number of coins possible in change, what would you expect them to be?

37. How many 10p are there in £5?

Fill in the blanks and make each of these into identical sets:

38. {5, 15, 25, 35}={25,   , 35, 5}

39. {P, L, N, M}={L,   , P, N}

40. {△, ○, □, △}={○, △, □,   }

Complete these series:

41. 2,   , 6, 8, 10.

42. 1, 4,   , 10, 13.

43. 1, 4,   , 16, 25.

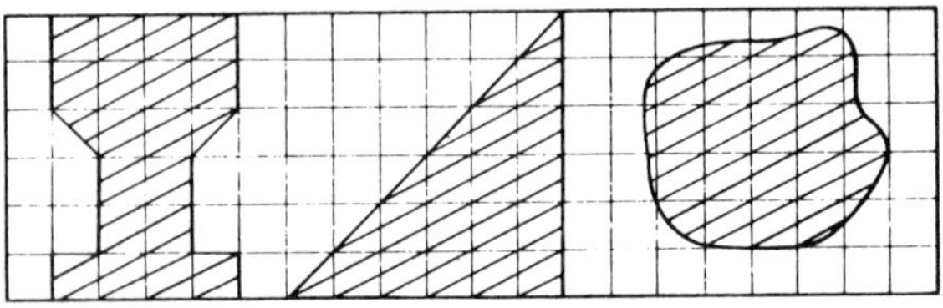

44. 45. 46.

Each square is a square centimetre. How many square centimetres are there in each?

In question 46, a square that has more than a half shaded counts as a full square, one which has less is ignored.

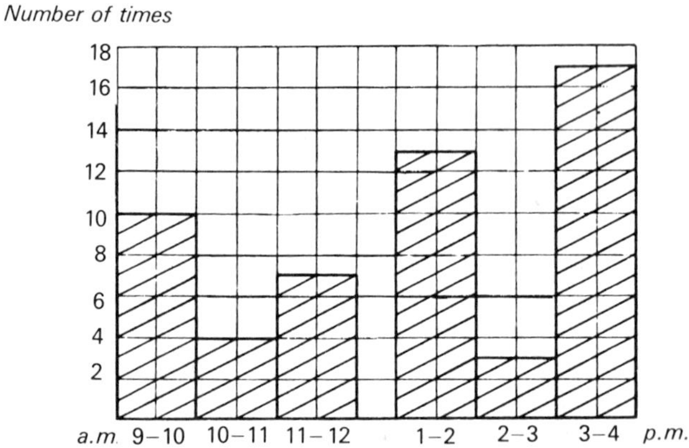

This graph shows the number of times the classroom door was opened during Monday.

47. How many times was it opened altogether?
48. During which hour was it opened most times?
49. What was the difference between the greatest and the least number of openings?
50. Was it opened most in the morning or the afternoon?
51. What was the average number of times an hour it was opened?
52. The British High Jump record is 2·80 m. for men and 1·95 m. for women. How much higher in cm. is the men's record than the women's?

This table shows distances from Paris on the N 6 route in France.

Paris to		How far is it from
Chalons	214 km.	53. Chalons to Avignon?
Lyon	293 km.	54. Lyon to Marseille?
Avignon	434 km.	55. Marseille to Chalons.
Marseille	493 km.	56. Driving at an average speed of 55 km./h. how long would it take to drive from Avignon to Chalons?

Fill in the gaps in this table:

	Car	Petrol used (Litres)	Distance covered (Km.)	Consumption (Km./Litre)
57.	APY114E	71	781	
58.	DCF77D	20	300	
59.	NET111C	123		12
60.	ALF774		4,563	13
61.	ZUX57H	177		10

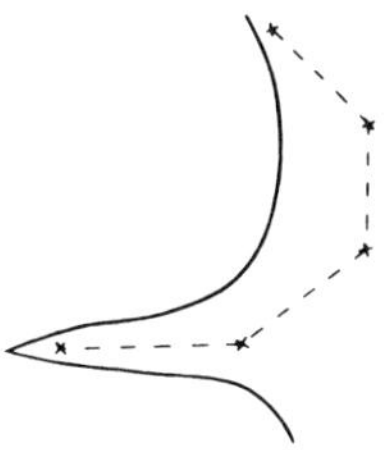

62–65. A ship leaving port changes course four times. What do you estimate each of these bearings to be?

Mathematics Test 5 *(40 minutes)*

1. Select two sub-sets from this list:
   Mr. Jones, Mrs. Jones, Tom, Freda, Helen.
2. What is the sum of ·25 and ·06?

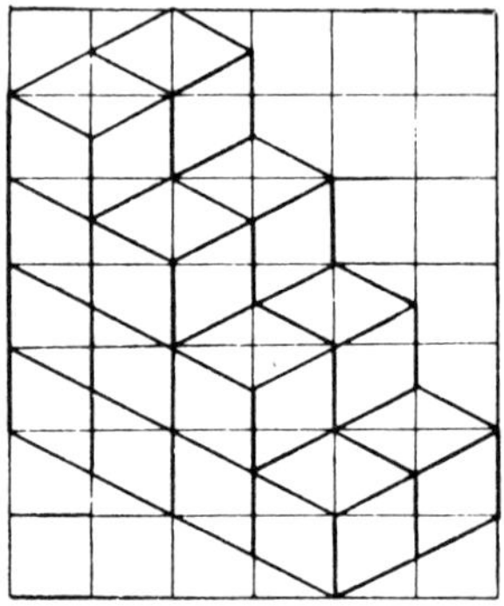

3. How many square blocks are there in this pile?
4. If the average contents of a box of matches is 50 and I count four boxes and find 45, 52, 47 and 41, how many matches must I find in the fifth box to keep the same average?
5. Write four tens, five units and three tenths as one number.
6. How many pence are three rabbits at £1·30 each?
7. What change would you receive for a £5 note if you bought 2 kg. of apples at 40p a kilogram and 500 g of nuts at £2.20 a kilogram?
8. A barrel holds 55 litres. If 27 half-litre bottles are filled from it, how much will remain?

This shows how many thousand people attended a football match and how much they each paid to go in.

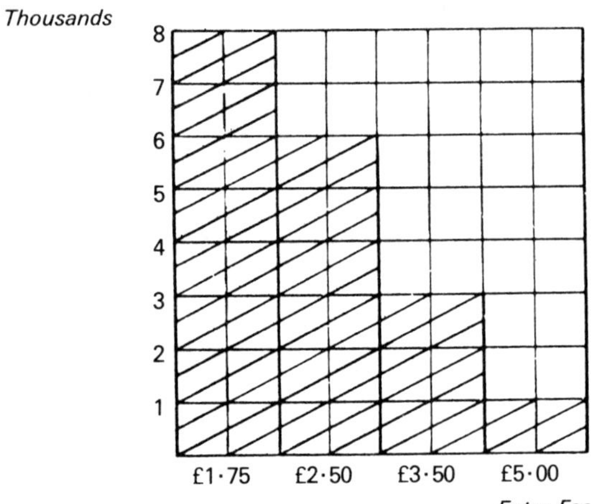

9. How many people attended in all?
10. How much money was taken in all?

Winchester	**London**
dep.	arr.
7·25	8·15
9·30	10·41
12·45	13·50
14·06	15·15
16·10	17·05
19·11	20·19

11. Which is the fastest train of the day?
12. Which is the slowest?
13. How many minutes does the journey take on the average?
14. If I want to be in London by 4 p.m., what is the latest train I can catch?
15. A jersey costs £11·50. The price is reduced by 10%. How much does it cost now?

In each equation the same sign has been omitted on both sides. What is it?

16. 6 3=10 5 17. 14 8=7 15 18. 25 3=32 10

19. Write 15 minutes as a percentage of an hour.
20. If a man walks at 6 km. an hour, how long does he take to walk 15 km.
21. If oranges are packed in a crate 6 to a row in 8 rows and there are 8 layers and they are repacked in boxes that hold a dozen each, how many boxes will be needed?

Class collection for Christmas Fund

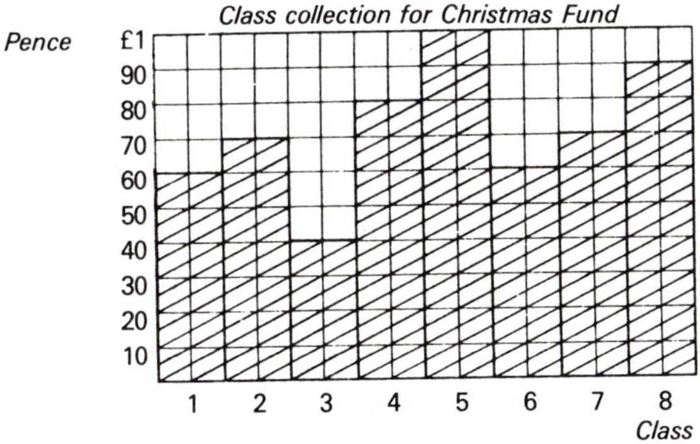

22. Which classes collected the same amount of money?

23. How much more did the class that collected the most get than the one that collected the least?

24. Which classes collected more than the average?

25. How much was collected in all?

The weights of the four members of a family are
Mother 52·4 kg., Father 67·5 kg., Tom 22·3 kg., Mary 18·2 kg.

26. What is the total weight of the family?

27. What is the average weight of these four people?

28. Copy this and shade the place where you would put the names of children who had hens and cats but no dogs.

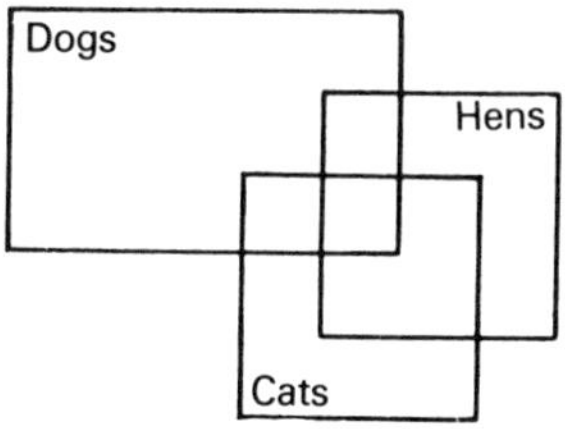

29. The readings at the beginning and end of an hour on a car's speedometer were 13,954 km., and 14,006 km. At what average speed in km./h. had the car been driven during this hour?

30. To what measurement must a pair of compasses be set to draw a circle 58 cm. in diameter?

31. If $\pi = 3 \cdot 14$, what is the circumference of a circle with a radius of 2 metres?

Complete the magic square:

32.	16	2	3	
33. 34.		11	10	
35.	9	7		12
36.	4	14		1

37. All the angles in this triangle are equal.
    How big is each?

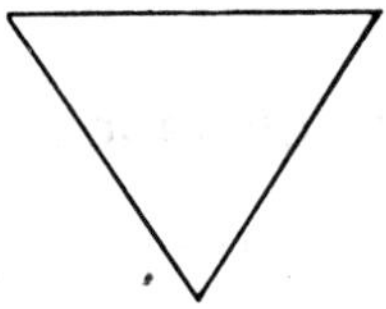

38. How many angles of this size could you
    fit into the centre of a circle?

39. How many axes of symmetry has this regular pentagon?

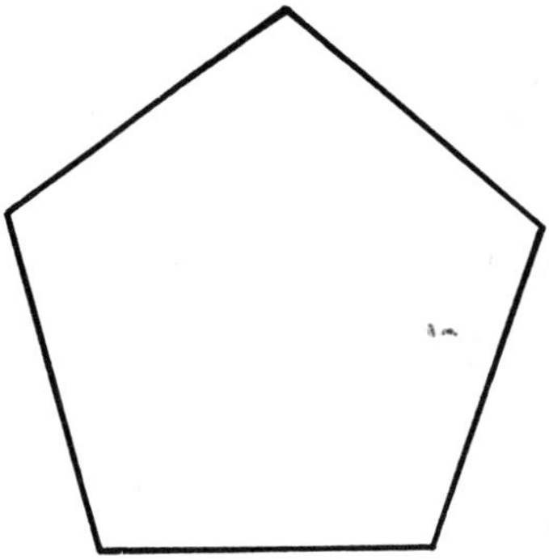

40. Change these numbers to the base 8.
    19, 99, 71.

41. In a number test of 100 items I had 22 wrong.
    What percentage had I correct?

42. Write this fraction as a decimal:
    $\frac{67}{100}$

43. Complete the following:
    5·5, 5·2, , , , , , 3·4.

44. My watch loses 4 minutes every hour. If I put it right at noon on Tuesday, what time will it show at 16·00 on Wednesday?

45. Express $2^2 \times 2^3$ to the base two.

Intelligence Test 1 *(25 minutes)*

A B C D E F G H I J K L M N O P Q R S T U V W X Y Z

1. How many letters are there after T?
2. What would be the thirteenth letter of the alphabet if C, F, and P were left out?
3. SMALL FACE THUNDER. Which of the first four letters of the alphabet does not occur in these three words?
4. Write the middle word in this sentence.
5. What is the fourth letter before P?
6. CANYON CRAYON What letter appears in the second word but not in the first?

These words are arranged in families, but there is one word in each line that does not belong to the family. Find it.

7. soldier : sailor : firework : airman : postman
8. cabbage : cod : potato : turnip : carrot
9. house : hut : bungalow : villa : ostrich
10. dog : horse : canary : cat : cow
11. coal : air : iron : brick : stone
12. primrose : pine : beech : oak : ash

Look at this.

Sea is to blue as (stay : grass : air) is to (green : mud : lark).

The sea is blue. In the brackets are the words "grass" and "green." These have been underlined because

Sea is to blue as grass is to green.

Now do these. You have to choose one word from each group in brackets.

13. Grass is to green as (cake : snow : bell) is to (fair : clear : white).

14. Cat is to kitten as (dog : veil : sound) is to (ape : puppy : tree).

15. Brick is to wall as (fat : fire : tile) is to (roof : train : arrow).

16. Brother is to sister as (men : uncle : clerk) is to (fairy : leaf : aunt).

17. Long is to short as (wide : hill : brief) is to (shelf : narrow : grid).

Smooth is to rough as (sleek : grin : art) is to (dear : anvil : rumpled).

19. Often is to seldom as (never : gain : many) is to (few : very : always).

20. Big is to little as (small : huge : elf) is to (tiny : all : smoke).

21. Copy out all the numbers in this line that will not divide by three. 5, 6, 15, 21, 8, 18, 36, 2.

Here as some rows of numbers. In each row the numbers are arranged in a special order. Find out what the order is and then write down two more numbers that could be added to each of these lines to carry on the order.

22. 1. 2. 3. 4.

23. 1. 4. 7. 10.

24. 3. 4. 6. 9.

25. 35. 30. 25. 20.

26. 12. 11. 9. 8.

27. 2. 5. 9. 14. 20.

28. 1. 4. 6. 9. 11. 14.

Each of these sets of words could be written in a special order. Choose the word that should be in the middle of each set.

29. big : biggest : bigger
30. three : five : four
31. warm : cold : hot
32. June : May : April
33. bottom : top : middle
34. sort : post : deliver
35. walk : dawdle : run
36. train : bicycle : bus

Choose the word that means the opposite, or almost the opposite, of the first word.

37. *black* yellow : white : crimson : green : blue
38. *high* up : fly : top : low : bottom
39. *north* east : point : pole : south : west
40. *back* front : side : arm : leg : head
41. *flying boat* yacht : submarine : battleship : ship : boat
42. *generous* clever : silly : mean : good : able
43. *sell* buy : gain : harvest : sew : blind

Choose the word that describes all the other words in the line.

44. family : husband : wife : son : daughter
45. beech : ash : tree : maple : elm
46. rain : sea : river : water : stream
47. wheat : cereal : oats : rye : barley
48. hake : cod : herring : salmon : fish
49. grocer : shop-keeper : draper : butcher : fishmonger
50. silk : cotton : wool : fibre : hemp

Intelligence Test 2 *(25 minutes)*

A B C D E F G H I J K L M N O P Q R S T V W X Y Z

1. What letter has been missed out of this alphabet?
2. What letter lies midway between the tenth and the fourteenth?
3. If *feline* contains more letters than *follows* write "get", if it contains less write "take."
4. *Barbarian.* Which letter in this word comes nearest the last letter of the alphabet?
5. *Contaminated.* Three letters appear twice in this word. Which pair lies between another pair?
6. *Surround.* What letter is repeated first in this word?
7. What letter comes before the letter but one after P?

Which of these words comes first in a dictionary?

8. gun : gut : gain : gone : grin : grip
9. fare : faint : fade : fair : fan : flap

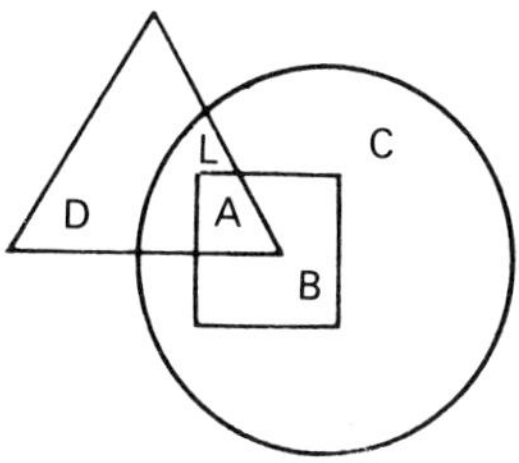

10. What letter is in the triangle only?
11. What letter is in the triangle and the circle but not in the square?
12. What letter is in triangle, circle, and square?

Think of the order in which these things should come and then write the first and the last.

13. second : hour : day : minute : week

14. May : July : April : June : August

15. trot : run : stand : walk : stroll

16. $\frac{3}{4}$: $\frac{10}{12}$: $\frac{1}{2}$: $\frac{2}{3}$: $\frac{5}{6}$

17. sergeant : private : captain : general : colonel

18. EBG : BBD : DBF : ABC : CBE

19. 0·0001 : 0·1 : 0·00001 : 0·01 : 0·001

Find a word that means the opposite to the first word in the line and has the number of letters shown in brackets. You must use the right number of letters.

20. *large* (5 letters)

21. *come* (2 letters)

22. *wide* (6 letters)

23. *green* (3 letters)

24. *nephew* (5 letters)

25. *rough* (6 letters)

From the words at the end of the line choose the one which goes best with the first three.

26. *apple pear orange* cabbage : peach : oak : beet

27. *liner yacht tug* cycle : train : trawler : car

28. *puppy kitten foal* dog : cat : horse : calf

29. *chief main leading* latter : principal : last : clan

30. *contract condense decrease* diminish : increase : extend : relate

Choose the words in each line that mean the opposite, or almost the opposite, of each other.

31. little : owl : big : dog : mule

32. canal : noisy : circus : film : silent

33. feel : float : run : sink : fall

34. deliver : collect : errand : fly : deny

35. happy : easy : idle : sad : simple

36. small : wealthy : sugar : poor : sweet

37. danger : complicate : dilute : dispute : concentrate

The numbers in each of these lines are arranged in a special order. Write down two more numbers that could be added to each line.

38. 5. 6. 7. 8.

39. 23. 20. 17. 14.

40. 17. 27. 37. 47.

41. $\frac{9}{10}$ $\frac{8}{20}$ $\frac{7}{30}$ $\frac{6}{40}$

42. 2. 4. 7. 9. 12. 14.

43. 1. 2. 2. 4. 3. 6.

Choose one word from those in brackets to complete the sentence.

44. Water is never . . .
(cold : dry : hot : wet : damp)

45. Animals always have . . .
(four legs : two arms : a tail : a head : fur)

46. A mountain is always . . .
(small : rocky : white : high : cold)

Which would be the middle word if these words were arranged in alphabetical order?

47. black : blue : blank : blame : blunt

48. knave : knight : knell : knife : know

49. If Tom is taller than Mary, and Mary is taller than Ann, who is the shortest?

50. How old is May, if Mary is two years older, and Robert, who is a year younger than Mary, will be ten next week?

Intelligence Test 3 *(25 minutes)*

Here are some codes and words:

H E L P	H A I L	F E W	E M P T I E D
1 3 5 9	1 7 8 5	10 3 19	3 20 9 12 8 3 2

Write these words in the same code.

1. H A D
2. F A D E
3. P E L T E D
4. D E A L T

Make a word from each of these groups of letters.

5. T A C
6. N E N P Y
7. T A R S T
8. L B T A E

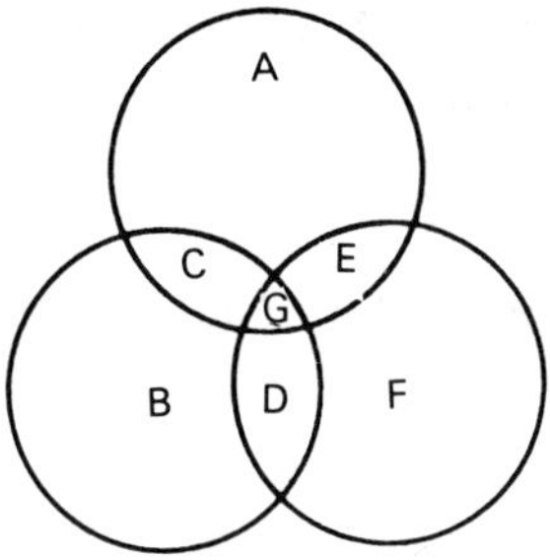

9. What letter is in the top circle only?
10. What letter is in all three circles?
11. What letter is in both the lower circles but not the top one?
12. What letter is in the top and left-hand circles but not in the right-hand one?

Select two terms in each line that do not fit in with the other four.

13. Monday : Tuesday : Easter : Friday : Christmas Day : Sunday
14. Scout : cat : Cub : Guide : man : Brownie
15. fall : swing : tumble : slip : flop : stand
16. ? : A : , : ! : 6 : ;
17. 4 : 12 : 6 : 3 : 8 : 7
18. If SULP means DAWN, what does SUP mean?
19. If PLANFEDM means FRIGHTEN, what does EFALEDDM mean?
20. If XALMPUZMO means BEAUTIFUL, what DOES ZMOZUO mean?

Choose the word that does not fit in with the others.

21. cage : zoo : camel : drifter : wolf
22. cliff : railway : crag : precipice : waterfall

23. January : December : November : February : June
24. bird : cat : butterfly : moth : bat
25. deny : accuse : disavow : repudiate : disown

Write two more numbers that could follow the figures given in each line.

26. 7. 6. 5. 4.
27. 2. 4. 6. 8.
28. 2. 5. 3. 10. 4. 15.
29. 7. 7. 6. 6. 5. 5.

5 6 7 8 9 10 11 12 13

30. Which is the middle number in this line?
31. Which is the middle number of those that divide by three?
32. Which number is double another number and over ten?
33. Which two numbers added together would make the last number?
34. If one were added to it, which number would be twice the third number in the line?

Four girls play games. Mary and Bessie play rounders; Ann and Ruth play hockey; Mary and Ruth swim; and Mary and Ann play tennis.

35. Who does not play rounders or swim?
36. Which girls play with Ruth?

Choose one word from each bracket to complete the statement.

37. Heavy is to light as (far : large : many) is to (small : lead : kite).
38. Walk is to run as (trot : fit : help) is to (send : bring : gallop).

39. Snow is to glacier as (grain : rain : herring) is to (lake : fish : wool)

40. Admit is to exclude as (refuse : pretend : agree) is to (follow : deny : send).

Choose two words in each set that mean the opposite, or nearly the opposite, of each other, and another word that means the same as one of them.

41. left : flat : seldom : sloping : level
42. cold : conclude : begin : contempt : finish
43. ugly : beautiful : beautify : huge : plain
44. control : brave : contempt : cowardly : courageous
45. thrice : never : frequently : twice : often

Find a word that means the same as the given word and has the number of letters given in brackets.

46. help (3)
47. fasten (3)
48. pleat (4)
49. left out (7)
50. rise (6)

Intelligence Test 4 *(25 minutes)*

Which word includes all the others in the line?

1. crowd : men : women : children : uncles : aunts
2. lawn : lily : garden : grass : rose : path
3. page : title : full-stop : line : comma : book
4. roof : tool : manager : factory : workmen : machine
5. church : village : grocer : bus-stop : school : people

One word in each of these sentences has its letters mixed up. Write the correct word.

6. The dog carried Fred's aotc.
7. One ngieenv there was a red sunset.
8. The great sweva broke over the pier.
9. He saw him standing by the rigbde.

Answer these questions using a word from those in brackets.

10. Which of these has every house?
    (radiator : roof : garden : gas stove : bell)
11. Which of these things is never made of glass?
    (window : spectacles : vase : chimney-pot : tumbler)
12. Which of these never flies?
    (mosquito : mountain : swan : sparrow : bird)
13. Which of these words describes all the others?
    (villa : farm : flat : hut : dwelling)

What letter occurs most times in the following sentence?

14. The blue bull wallowed in the muddy pool.

George	Mary	Bill
June	Tom	Jill
Harry	Joan	Ted

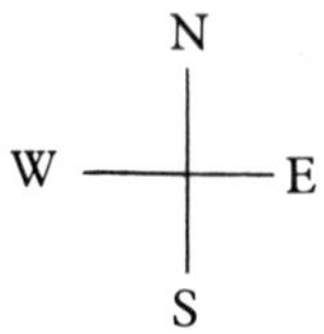

Here are some children standing in a square.

15. What is the name of the girl in the north side?
16. Who is standing in the centre?
17. Who can face south and east without anybody being in front?

18. Which girl is farthest west?

19. What boy is standing due south of Bill?

20. If they all face north, who is on the right of Tom?

21. If they turn to face south again, who is then on the right of Ted?

Finish these statements by selecting one word or number from each group in brackets.

22. Cub is to wolf as (calf : breakfast : hen) is to (fox : cat : cow)

23. Room is to hotel as (cage : shelf : stove) is to (barracks : zoo : cinema)

24. Tree is to forest as (cell : pupil : kennel) is to (dog : house : class)

25. 4 is to 12 as (15 : 11 : 0) is to (6 : 45 : 71)

26. Elevate is to depress as (end : finish : lift) is to (lower : upper : side)

Arrange these words in the right order to make a sentence. Write down the middle word of the sentence.

27. Tom away ran

28. poured the window down rain

29. blue bright sky was the

30. large elephants tails the were holding fourteen each other's.

Write down the two numbers that have been left out.

31. 8 9 __ 11 12 __

32. 3 __ 9 12 15 __

33. __ 15 20 25 __ 35

34. 14 25 36 47 __ __

35. 0·73 __ 73 730 7300 __

Mary has a red ribbon. She gives this to Jill in exchange for a blue one which Jill has had from Maud for a yellow one. Maud bought the blue one from Vera for two pence. Dora's mother will not let her part with her white one.

36. Who has the blue one now?
37. Who has not changed her ribbon?
38. Who has no ribbon to finish with?
39. Who had the blue ribbon at first?
40. Which girls exchange their ribbons twice?

Choose the right word from each group in brackets to complete these sentences.

41. (red : blue : black) is to danger as green is to safety.
42. (kg. : g. : mg.) is to weight as mm. is to length.
43. House is to room as (many : people : room) is to one.
44. Agile is to (land : clumsy : quick) as fast is to slow.

Choose the word that means the opposite, or almost the opposite, of the one at the beginning of the line.

45. *muddy* plain : muddled : puddle : stream : clear
46. *hinder* persist : consist : assist : construct : persuade
47. *expand* impress : contact : deduce : contract : explain
48. If April 9 is on a Friday, what is the second letter of the day of the week on which April 18th falls?

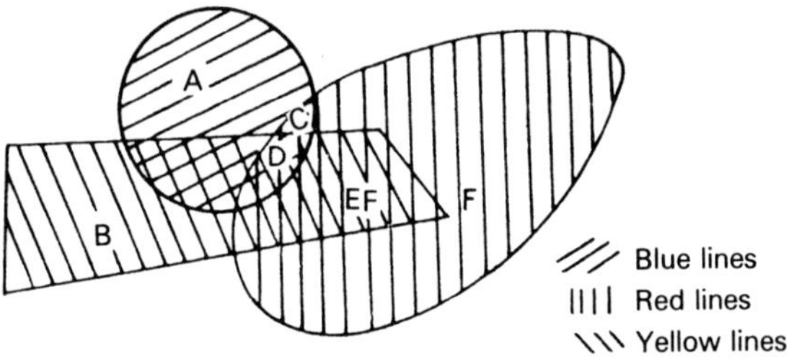

49. What part is coloured blue and red?
50. What part has all colours?

Intelligence Test 5 *(25 minutes)*

Copy out these sums, putting in plus and minus signs to give the right answer.

1. 1 2 3=6
2. 2 2 4=4
3. 5 3 7=9
4. 6 3 2=7
5. 7 11 8=26

Jane and Mary have red ribbons.
Mary and Sally have blue ribbons.
Maud and Jane have green ribbons.
Jane and Susan have yellow ribbons.

6. Who has most ribbons?
7. Who has only one ribbon and that a blue one?
8. Who has a red ribbon but not a yellow one?
9. Who has both a red and a blue ribbon?
10. Who has a red and a green ribbon but not a blue?

Choose the three that are alike in each of the following

11. cm. : ft. : m. : in. : mm.
12. ash : hen : elm : man : rose : oak
13. home : pond : stream : film : sea : garden
14. cycle : elephant : ant : bus : chimney : car
15. fish : hat : coat : sword : helmet : cap

This is how the children spent their holiday:

John, Mary, and Fred went to the sea.
Sally, John, and Robert went to the country.
Fred, Robert, and Mary went camping.

16. Which boy went camping by the sea?
17. Who went to the country and the sea?
18. Who went to the sea but did not camp?
19. Which boy went to the country but did not camp?
20. Who camped in the country?

Choose one word from each group in brackets to complete these sentences.

21. Full is to empty as all is to . . .
    (some : none : most : tell)
22. Letter is to envelope as foot is to . . .
    (shop : feet : shoe : box)
23. Longer is to shorter as week is to . . .
    (big : day : close : near)
24. Dog is to kennel as lion is to . . .
    (jungle : flat : den : tree)
25. Goldfish is to bowl as swimmer is to . . .
    (sea : river : sky : pond)

Write down figures to fill in the gaps to complete each series of numbers.

26. 10 __ 30 __ 50 60
27. __ __ 10 12 14 16
28. 5 __ __ 20 25 30
29. 96 48 24 12 __ __
30. 1 4 2 8 3 12 __ __

Choose two words from each group in brackets that will complete these sentences.

31. Fish is to sea as . . . is to . . .
    (bird : water : lake : air : fire)

32. First is to last as . . . is to . . .
    (end : once : beginning : send : come)

33. Horn is to car as . . . is to . . .
    (train : bell : bicycle : horn : lamp)

34. Clean is to dirty as . . . is to . . .
    (shirt : laundered : sky : soiled : touch)

35. Tick is to clock as . . . is to . . .
    (water : crane : hum : click : dynamo)

Select one word in each line that does not fit in with the others.

36. ten : eleventh : three : five : eight

37. black : blue : brown : blink : bland

38. flock : bird : herd : swarm : shoal

39. stand : sit : lie : squat : walk

40. strawberry : red-currant : raspberry : gooseberry : banana

Choose the number that is most unlike the others.

41. 2 4 5 8 12

42. 4 6 12 16 8

43. 45 23 34 56 37

44. 49 35 7 25 63

45. 3 9 7 11 17

Write the figures that should go in the blank squares to complete these patterns.

46.

1	2	3
8	9	4
7	6	

47.

9	8	7
10	9	8
	10	9

48.

3A	2A	1A
3B	2B	1B
3C	2C	

49.

5	2	7
4	1	5
9	3	

50.

2	4	8
3	9	27
4	16	

Intelligence Test 6 *(25 minutes)*

Each of these words has a letter in front of it. Pick out the word from each line that is different from all the other words and copy the letter in front of it.

1. (*a*) cow (*b*) robin (*c*) dog (*d*) cat
2. (*a*) painter (*b*) plumber (*c*) cobbler (*d*) bricklayer
3. (*a*) coke (*b*) coal (*c*) wood (*d*) slate
4. (*a*) woman (*b*) man (*c*) lady (*d*) girl
5. (*a*) carpet (*b*) linoleum (*c*) wallpaper (*d*) rug

From the words with letters in front of them pick out one which is the opposite, or almost the opposite, of the word at the beginning of the line.

Write down the letter which is in front of the word you select.

6. *sea* (*a*) ocean (*b*) river (*c*) land (*d*) pond
7. *huge* (*a*) big (*b*) wee (*c*) great (*d*) massive
8. *flowing* (*a*) running (*b*) ripple (*c*) stagnant (*d*) clear
9. *stupid* (*a*) sensible (*b*) unwise (*c*) remarkable (*d*) silly
10. *lying* (*a*) truthful (*b*) untruth (*c*) trustworthy (*d*) trusting

Give the letter in front of the word that includes all the other words.

11. (*a*) man (*b*) mankind (*c*) child (*d*) woman
12. (*a*) bush (*b*) laurel (*c*) rose (*d*) lilac
13. (*a*) tiger (*b*) leopard (*c*) carnivore (*d*) lion
14. (*a*) swede (*b*) hay (*c*) fodder (*d*) clover
15. (*a*) lemonade (*b*) tea (*c*) water (*d*) drink

Put these words into their correct order and then give the letters which are in front of the first and the last words in each group.

16. (*a*) dinner (*b*) tea (*c*) supper (*d*) breakfast
17. (*a*) vest (*b*) overcoat (*c*) shirt (*d*) jacket
18. (*a*) win (*b*) go (*c*) swim (*d*) dive
19. (*a*) M (*b*) N (*c*) L (*d*) O
20. (*a*) 5·1 (*b*) 4·55 (*c*) 5·5 (*d*) 4·59

Here are several series of letters that form patterns. Write the pairs (or groups) of letters that should go in the spaces to complete the patterns.

21. LL __ NN __ PP QQ
22. CD EF GH __ __
23. __ __ CX DW EV
24. LX __ __ OX PX
25. DEF FGH HIJ __ __

Tom was top in Arithmetic and second in Reading and Writing and Mary was top in Writing and third in Reading. Jill was second in Arithmetic and third in Writing. Frank was top in Reading and second in Arithmetic.

26. Who was top in writing?
27. Who had most second places?
28. In which subject was there a tie?
29. If First counts 4 points, Second 2, and Third 1, who had most points?
30. Who beat Mary in Reading?

	Friday	Saturday	Sunday
Wet	10	5	2
Fine	2	5	8
Dull	1	3	2

This table shows the number of wet, fine, and dull Fridays, Saturdays, and Sundays in the summer.

31. How many Saturdays were fine?
32. How many more Sundays than Fridays were fine?
33. How many more wet than fine days were there?
34. On which day would there have been most chance of fine weather?
35. Under which heading did least days come?

Write the figures or letters that should be placed in the blank squares to complete these patterns.

36.

1	2	3
4	5	6
7	8	

37.

2	4	6
4	2	4
6	4	

38.

1	2	3
2	4	6
3	6	

39.

A	B	C
C	E	G
E	H	

40.

A		A
B	2	B
C	3	C

These words are arranged in patterns.

Thus in FADE FAD you see the E has been left off FADE to make FAD, so from MADE you can make MAD.

FADE FAD MADE MAD

Find the patterns and complete these groups of words.

41. SAME SAM CAME
42. CARRY CAR EARTH
43. MANE NAME MADE
44. WINDOW WIND BINDER
45. CAME COME MIND

Twenty-seven match-boxes are stuck together to make a shape that is three boxes high, three boxes wide, and three boxes deep. The top is then painted red, the bottom green, and the sides yellow. (A match-box has six sides.)

46. How many boxes have one side red?
47. How many boxes are not painted at all?
48. How many have both green and yellow on them?
49. How many have two sides yellow and one red?
50. How many have yellow on one side and no other side coloured?

Intelligence Test 7 *(50 minutes)*

A B C D E F G H I J K L M N O P Q R S T U V W X Y Z

1. Which letter in FRAIL comes nearest to the letter T?
2. Which letter comes next after K in the alphabet without being in CLIMAX?
3. What letter comes more than once in both SENSATIONAL and ANIMATES but not in CONVOLVULUS?
4. What would CONTUSION look like printed backward?
5. What letter is found most times in SEQUESTRATED?
6. What is the third letter of the last word in this sentence?
7. What letter comes midway between E and O if J and K are removed?
8. If A and then every third letter in the alphabet were taken away, what would be the thirteenth letter?

Choose the word from those in small letters that best describes the three in capital letters.

9. CAT COW ANTELOPE
   fowl : animal : herd : male : fish

10. SPEAR GUN ARROW
tool : shoot : man : weapon : crowd

11. STREAM POND RIVULET
river : lake : sea : water : run

12. BANG WHISPER SHOUT
sound : speech : tell : count : collect

13. GRAM LITRE METRE
length : weight : measure : rule : time

14. TRAIN BUS SHIP
journey : road : rail : covey : transport

15. DEN BURROW LAIR
house : home : animal : cage : eggs

16. ROSE THISTLE SHAMROCK
flower : seed : tree : rock : emblem

What two extra pairs of letters should be added to each of these lines to complete the patterns?

17. LM MN NO

18. AO BO CO

19. VV UU TT

20. PQ RS TU

21. RQ PO NM

22. ABC BCD CDE

Write down the pairs of numbers that could be added at the end of each line to complete the pattern.

23. 17 16 15 14

24. 13 15 17 19

25. 4 6 9 13

26. $\frac{1}{4}$ $\frac{2}{5}$ $\frac{3}{6}$ $\frac{4}{7}$

27. 4 3 5 4 6 5 7

28. 123 134 145 156

29. 1 4 8 11 15 18 22

Choose one word from each group in brackets to complete the following sentences.

30. End is to beginning as (front : back : last) is to (first : opening : stop).

31. Paper is to book as (wood : egg : shell) is to (window : desk : car).

32. Up is to down as (top : flat : ceiling) is to (end : tile : floor).

33. Cow is to milk as (hen : calf : sow) is to (ewe : egg : mustard).

34. Sweet is to sour as (flower : scent : sugar) is to (vinegar : bun : wheat).

35. Teacher is to children as (man : captain : soldier) is to (sailor : crew : meat).

Find words which are the opposite of the words that are given and have the number of letters shown.

36. *asleep* (5 letters)

37. *light* (4 letters)

38. *noisy* (5 letters)

39. *break* (4 letters)

40. *lift* (5 letters)

41. If *flimsy* is written *kpnsul*, how is *film* written?

42. If *famous* is written *splony*, how is *sam* written?

43. If *agriculture* is written *balnpoxcolz*, how is *calculate* written?

44. If *cultivation* is written *plmsvnasvbx*, how is *national* written?

Each of these sets can be arranged in a special order. If this is done, which will be the middle word or number in each set?

45. week : second : hour : day : minute
46. house : town : village : city : room
47. 486 : 684 : 864 : 648 : 468
48. oven : farm : market : shop : dinner-table
49. hurricane : breeze : calm : wind : gale
50. battleship : cruiser : yacht : destroyer : canoe.

Fill in the figures that have been left out of these sums.

51. $1? \times 7 = 77$

52. $9\overline{)45}$
 $\quad ?$

53. $3 + ? - 2 = 5$

54. $48 \div ? = 12$

55. $\frac{73 - ?}{5} = 12$

56. If the alphabet were written backward, which two letters would be moved least?

57. If all capital letters that can be formed of only straight lines were removed from the alphabet, which would be sixth from the end?

Choose one word from each group in brackets to complete each statement.

58. Far is to near as (way : long : measure) is to (big : strong short).

59. Fish is to (river : eat : catch) as monkey is to (banana : tree : bird).

60. (high : hill : snow) is to mountain as coverlet is to (bed : eiderdown : sheet)

61. (spring : leg : cushion) is to chair as (pillow : bedroom : window) is to bed.

62. Fish is to (sea : water : scale) as hen is to (feather : egg : nest).

63. Station is to train as (airport : engine : pilot) is to (car : ship : aeroplane).

64. Seek is to (send : find : fend) as (hide : lend : wait) is to seek.

65. (animal : girl : hen) is to (boy : child : friend) as cow is to bull.

66. Try is to (gain : lose : well) as (win : endeavour : forget) is to win.

67. Class is to school as (room : boy : hall) is to (girl : class : headmaster).

Choose the one term in the line that includes all the others.

68. hockey : football : game : darts : rounders

69. little : huge : small : enormous : size

70. piston : sparking-plug : cam-shaft : engine : cylinder

71. cows : herd : cattle : buffaloes : elephants

72. Arithmetic : Geography : subject : History : English

73. decimal : digit : fraction : number : ten

74. column : newspaper : paragraph : headline : word

75. week : hour : month : year : day

76. lamp : pavement : gutter : pillar-box : street

77. frequency : often : seldom : never : frequently

78. lamb : mutton : ewe : ram : sheep

79. C : V : L : I : IV

These are word families. Choose the one word in each line that does not belong to the family.

80. write : speak : whisper : shout : exclaim
81. chair : furniture : field : table : stool
82. pen : needle : typewriter : pencil : crayon
83. beet : turnip : parsnip : cabbage : carrot
84. smooth : rugged : flat : plain : level
85. beseech : ask : beg : say : entreat
86. reap : harvest : sow : gather : collect

Choose two words in each line that mean the opposite, or nearly the opposite, of each other.

87. angry : huge : queer : tiny : soft : single
88. foiled : forbid : foretell : allow : forecast : end
89. arrive : foiled : entry : follow : depart : send
90. except : bold : confirm : feed : fasten : deny
91 complete : please : add : begin : submit : annoy
92. finish : follow : promote : permit : prevent : secure
93. signify : frustrate : fight : tighten : calm : relax
94. transport : follow : send : proceed : precede : placate
95. boisterous : transpose : belligerent : tranquil : present : follow
96. edit : regret : release : cancel : reform : arrest

If these words or numbers were arranged in order, which would be the first and the last in each group?

97. grandfather : baby : man : boy : youth
98. tram : cycle : car : bus : aeroplane
99. dusky : white : grey : cream : black
100. 2,200 g. : 1,010 g. : 3 kg. : 1 kg. : 999 g.

Intelligence Test 8

Choose the shape on the right of the line which is most like the three shapes on the left of the line, and copy the letter below it.

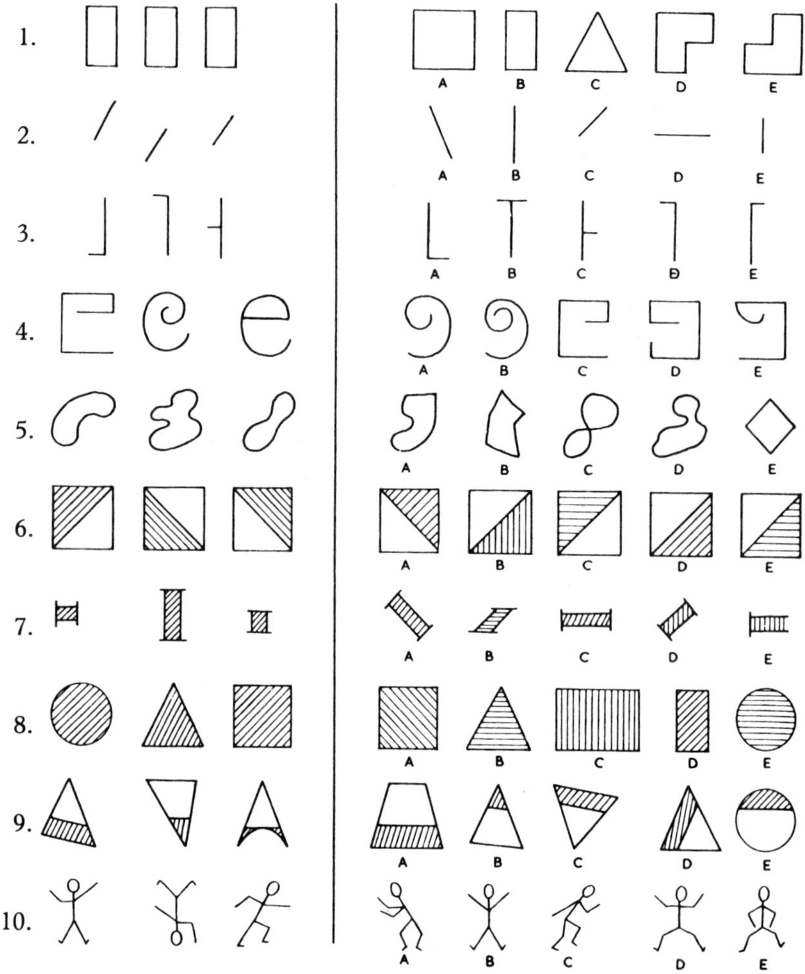

Select the square on the right of the line which you would take to fill the space on the left of the line, in order to make these shapes into a set; copy the letter below it.

11. 

12. 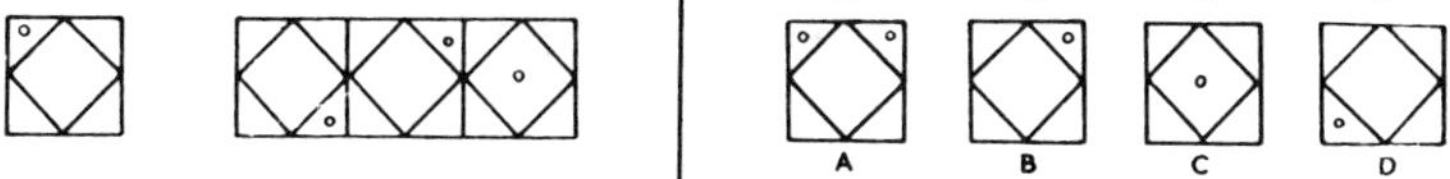

13. 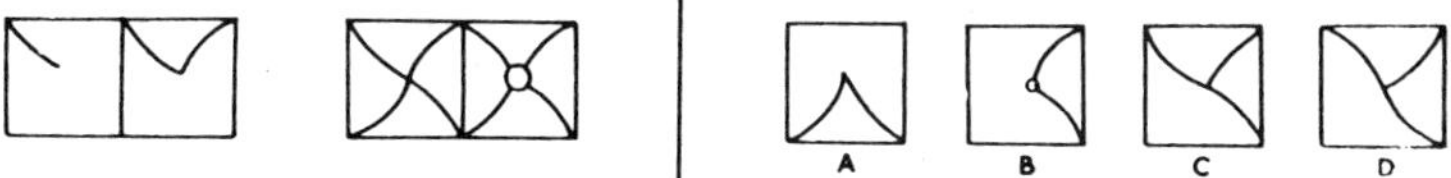

14. 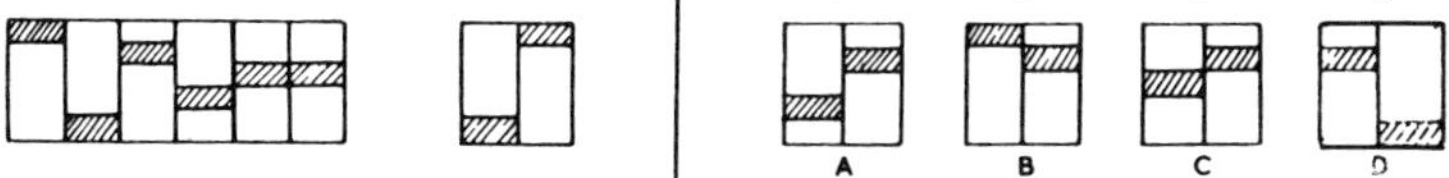

15. 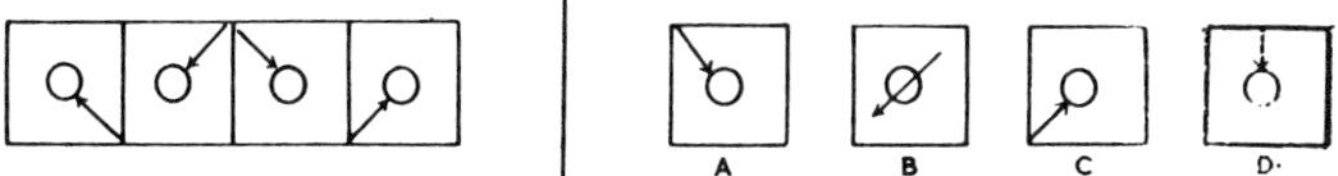

16. 

17. 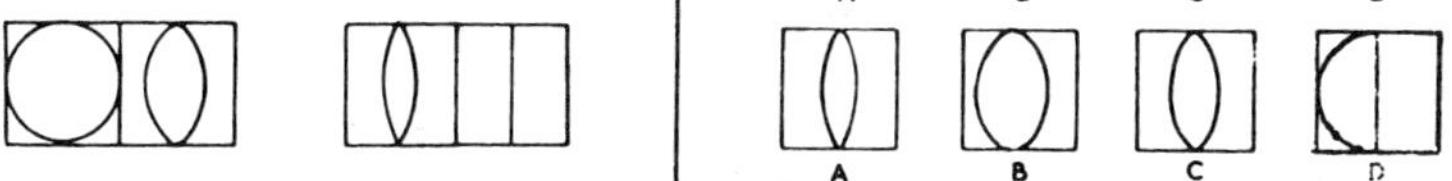

18. 

19. 

20. 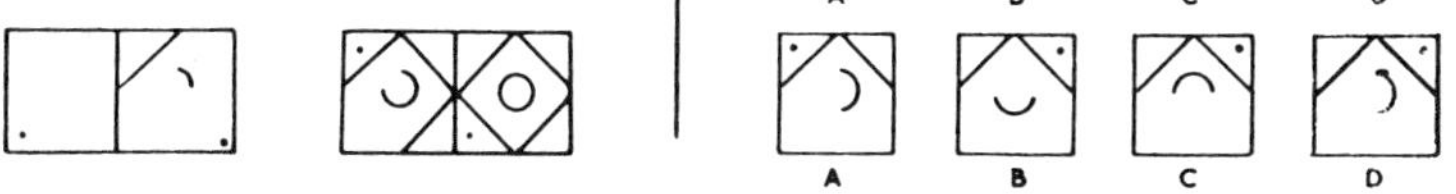

Select the shape from the group on the right that will complete the statement on the left; copy the letter below it.

21. 

22. 

23. 

24. 

25. 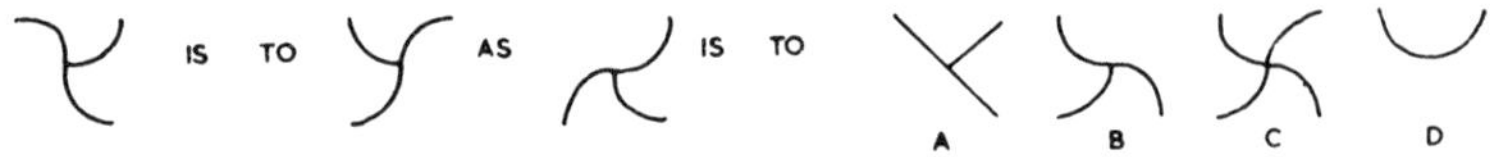

26. 

27. 

28. 

29. 

30. 

Select the shape in each line which is least like the other shapes in that line and copy the letter below it.

31. 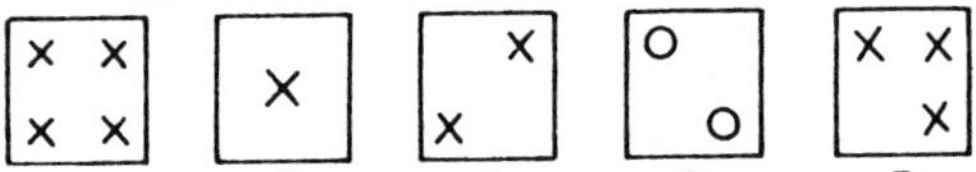

A B C D E

32. 

A B C D E

33. 

A B C D E

34. 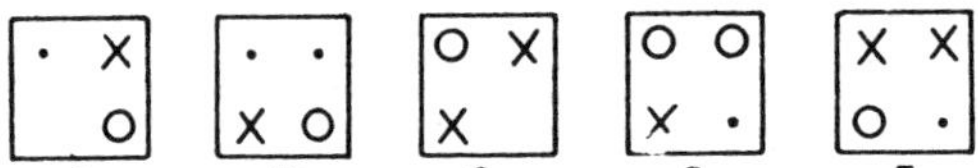

A B C D E

35. 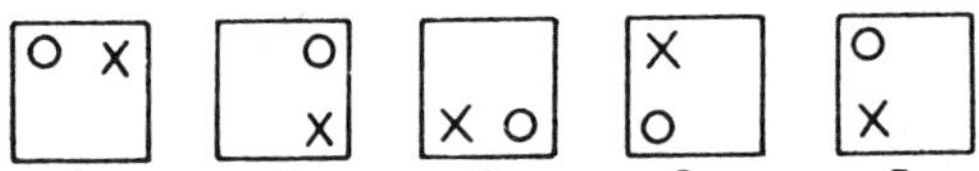

A B C D E

36. 

A B C D E

37. 

A B C D E

38. 

A B C D E

39. 

A B C D E

40. 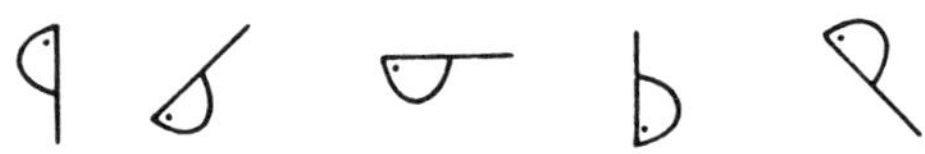

A B C D E

If these shapes were arranged in their logical order, choose the one which would come in the middle; copy the letter below it.

41.

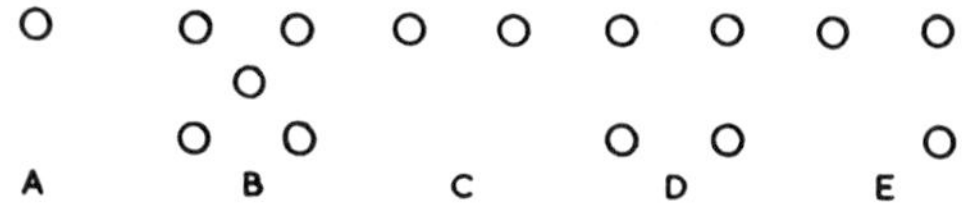

42.

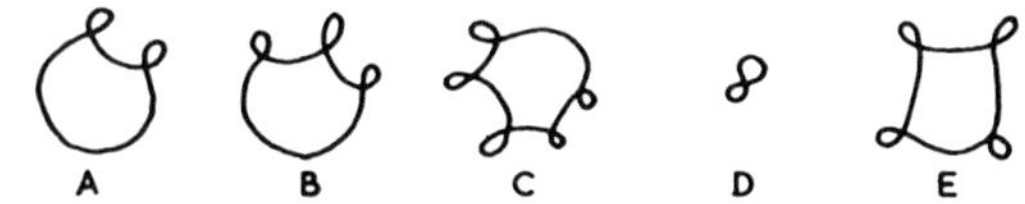

43.

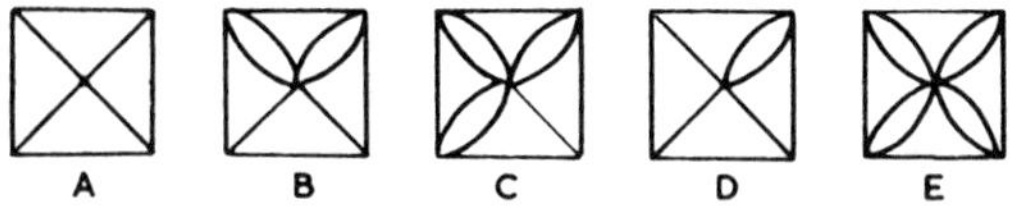

44.

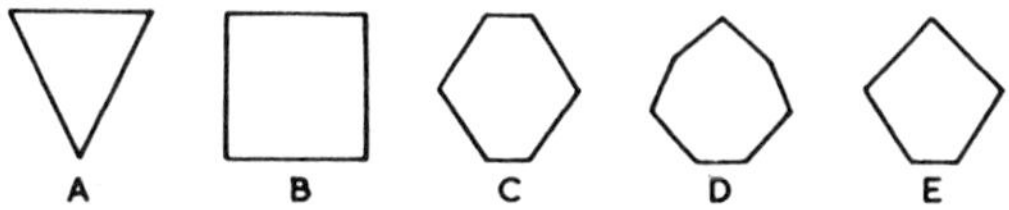

45.

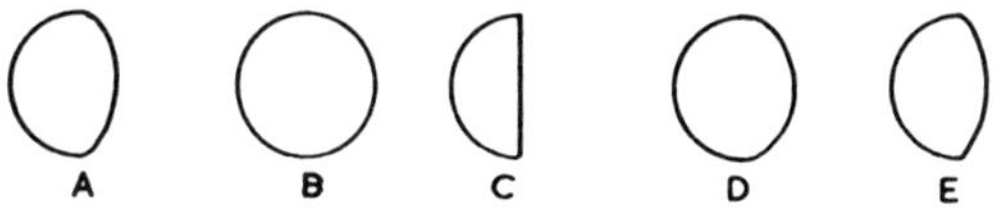

46.

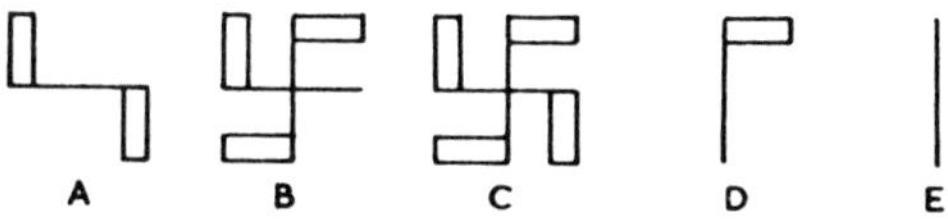

47.

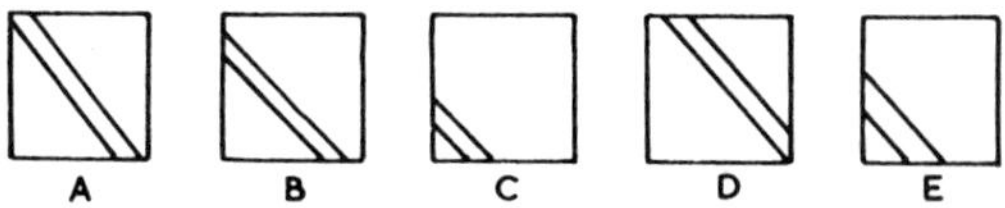

48.

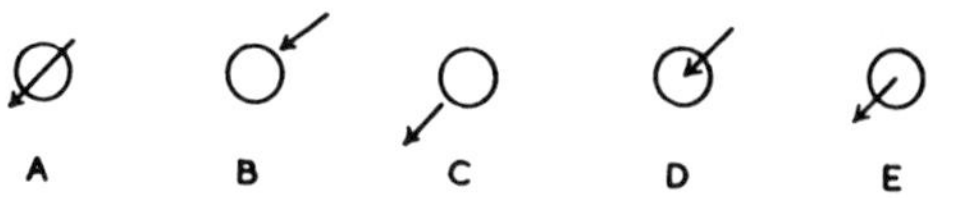

49.

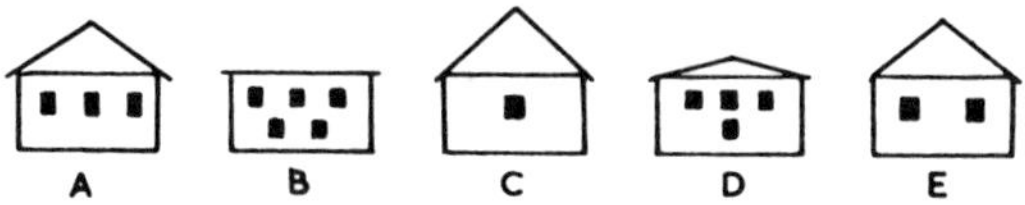

50.

Answers

English Test 1

1. the wind had dropped
2. Fred
3. shingle
4. above high-tide mark
5. they were going into a cave
6. running
7. cautiously
8. trouble
9. decided how he would do it
10. his sister
11. he cut them out
12. the ceiling
13. he pulled out the middle of a roll
14. silver and gold
15. dining-room
16. cards
17. Mount Everest
18. his own kit
19. twisted canes
20. the porters
21. noon
22. they swung
23. on narrow paths
24. in single file
25. wheat
26. the chaff
27. the small seeds
28. carries the cut corn into the machine
29. the beater
30. the sieves
31. a lorry

English Test 2

1. injure
2. permit
3. for example
4. that is
5. feed
6. obstinate
7. here
8. send
9. well
10. were
11. neither nor
12. quietly
13. complete
14. raise
15. valiant
16. disappear
17. regulate
18. prophesy
19. butcher
20. plumber
21. pilot
22. conductor or conductress
23. tobacconist
24. cutting
25. blackest
26. carried or carries
27. made
28. daily
29. asking or asked
30. business
31. dependable
32. O I
33. E I
34. E A
35. E
36. A I
37. I N U I
38. fleet compete
39. flew flue
40. retail tale
41. one fun
42. see flea
43. very apt
44. thoroughly reliable
45. a close similarity
46. reason
47. haughty
48. Tom asked, "Can I go now?"
49. The *Guardian* is a daily paper.
50. "Can you tell me," said Mary, "if he will come?"

English Test 3

1. Bob
2. a lake
3. along a sunken ledge
4. Tim
5. he flashed a light
6. a rope
7. four
8. a chest
9. he slipped out of his collar
10. the Professor
11. Bennet calmed him down
12. a wolfhound
13. we heard the chain fall
14. the Professor
15. a stupid invention
16. turn a wheel
17. a leather band
18. he decides to run away
19. on stronger plants
20. its red fruit
21. its fruits are poisonous
22. woody nightshade
23. purple and spotted
24. black and large
25. deadly nightshade
26. purple flowers in clusters and small berries

English Test 4

1. porter
2. shepherd or shepherdess
3. jockey
4. cobbler
5. pilot
6. sheet
7. brass
8. crystal
9. monkey
10. dust
11. their
12. wear
13. bigger
14. I were
15. he was a coward
16. to keep quiet about it
17. to make peace
18. politeness
19. strange
20. display
21. argument
22. fancy
23. plentiful
24. necessary
25. brief
26. persuaded
27. wintry
28. survivor
29. reduction
30. assistance
31. objective
32. We want to go to our homes
33. Some books fell from the tables
34. The boys were climbing over the walls
35. The horses jump the fences
36. abdicated
37. smithy
38. ignorance
39. ferry
40. opaque
41. ironmonger
42. navigator
43. perjury
44. seldom
45. curious

46. encouraged
47. impression
48. He shouted "Help! Help! I am drowning."
49. "Halt!" cried the sentry. "Who goes there?"
50. Mary and Fred, with all speed, flung up the window

English Test 5

1. an ape
2. at a farm
3. beech
4. they crept
5. open fields
6. tea-time
7. hunger
8. they were afraid
9. autumn
10. a plane
11. his surroundings
12. my companion was cheerful
13. the last leaves were falling
14. Romans
15. in garrison towns
16. 1 m.
17. south
18. to keep out the Picts
19. trimmed
20. the Solway Firth
21. to signal from
22. Sun
23. Moon
24. three heavenly bodies are in line
25. the sun goes behind the moon
26. Mars is so far away
27. the earth passing between the sun and the moon

English Test 6

1. (*a*)
2. (*c*)
3. (*c*)
4. (*b*)
5. (*a*)
6. (*b*)
7. (*c*)
8. (*d*)
9. (*d*)
10. (*c*)
11. take unfair advantage
12. I will have no more to do with it
13. be absolutely right
14. talk round the matter
15. try to do too much
16. make a mistake
17. be unable to make up your mind
18. cautiously
19. reafforested
20. enlisted
21. pretentious
22. denuded
23. (*a*) (*g*)
24. (*a*) (*f*)
25. (*c*) (*e*)
26. he did it
27. of men enjoy gardening
28. cycle more quickly than he can run
29. his fall was running too fast
30. of butter over margarine for cooking is undoubted
31. I wanted you to write carefully
32. Where did you go then?

33. I left you then to see what you would do
34. I came as soon as I could
35. Many great animals were crossing the river
36. lacking finish
37. not allowed
38. self-satisfied
39. waging war
40. outspoken
41. informer
42. completion
43. rounding
44. forceful
45. contemptuous
46. "Where do you come from?" inquired Lily.
47. "Is it a treacle well?" asked Alice.
48. "I feel most unwell," said the King.
49. "Come on," shouted the crowd. "Shoot!"
50. "Of course I will," I said, "if you will let me."

Arithmetic Test 1

1. 1009
2. 2691
3. 3332
4. 613 rem. 6
5. 7 m. 2 cm. 5 mm.
6. 7·2608 kg.
7. £3.19
8. 10·63 t.
9. £26.25
10. 12·21 l
11. 1 hr. 35 min. 54 sec. rem. 1 sec.
12. £5.90
13. 18 hr. 1 min. 20 sec.
14. £22.75
15. 3·049 kg.
16. 8 hr. 10 min. 56 sec.
17. 50·41127.
18. 81½p
19. 1·5215 kg.
20. 190 ml.
21. 100·755 kg.
22. 704
23. £2.05
24. 2 weeks 3 days 21 hr.
25. 6 hr. 9 min. 29 sec. rem. 3 sec.
26. 3,209 kg.
27. £40·25
28. 40·5 cm.
29. 42·292 km
30. 192 ml.

Arithmetic Test 2

1. 5
2. 50 p
3. 120 words
4. 3p
5. 250 kg.
6. 250 g.
7. 16
8. 10
9. 1·2 m.
10. 75 g.
11. £3.32½
12. 26p
13. 1/5
14. £1·90
15. 312
16. 10
17. 8
18. 8 pieces
19. 22 marbles
20. 56
21. 69 days
22. 9/20
23. £2.30 each
24. 100 trees
25. 2 tonnes
26. 400 km.
27. 17 boxes

28. 30 bottles
29. 60 lambs
30. 56 panels
31. £66
32. 12 teams
33. 2 hr. 36 min.
34. £35
35. 85 tonnes
36. $\frac{3}{8}$
37. £2.10
38. 80 times
39. $\frac{6}{7}$ kg.
40. $2\frac{1}{2}$ km.
41. £167·40
42. 12 cm.
43. 37
44. 150 people
45. 9 litres

Mathematics Test 1

1. 88732
2. 23788
3. 14
4. 606 books
5. 3 books
6. 5p
7. Waterloo
8. 15,000 passengers
9. 150,000 passengers
10. 500 trains
11. 284 trains
12. 2·43 kg.
13. 0·5
14. $\frac{1}{4}$
15. 16%
16. 0·56
17. $\frac{11}{25}$
18. 6
19. 24
20. 96
21. 20 24
22. 11 16
23. 150 m.
24. 42p
25. $\frac{1}{12}$, $\frac{1}{2}$, $\frac{4}{7}$, $\frac{3}{5}$, $\frac{9}{10}$
26. 5 m.
27. $1 \cdot 25 m.^2$
28. 4·9 m.
29. 25
30. 90°

Mathematics Test 2

1. 27·6 m.
2. 136 g.
3. 1,400 ml.
4. 3
5. 2
6. ×
7. ×
8. 345
9. 20 km.
10. Trident
11. 200 km./h.
12. 4,100 km.
13. 100
14. 5,268
15. 6
16. 19
17. £9
18. £400
19. Tuesday
20. £2,010
21. £335
22. $1 \cdot 9 = 1\frac{9}{10}$
    $1 \cdot 75 = 1\frac{3}{4}$
    $1 \cdot 2 = 1\frac{1}{5}$
23. 60 slabs
24. 3,996
25. 4
26. 24 km.
27. 3 km.
28. 21 km.
29. 30 boys
30. Any square covering four small squares

Mathematics Test 3

1. 84·994 kg.
2. 6·499 m.
3. 30 sec.
4. + or ×
5. ×
6. +
7. ×
8. ×
9. +
10. 18
11. 75 km.
12. 4 lights
13. 200 kg.
14. 57×42
15. 17
16. $\frac{177}{2}$
17. £54,750
18. 56 packets

19. 5,365 rem. 6
20. 204,036
21. 453
22. 1,812
23. 9,513
24. 906
25. 7.40 a.m.
26. 2.50 p.m.
27. 97 hr. 13 min.
28. 28 hr. 3 min.
29. 16 days 8 hr. 15 min.
30. 136·5327 t.
31. 4·895 l.
32. £7.47
33. 700 kg.
34. 4
35. 1/10
36. £8.40

Mathematics Test 4

1.

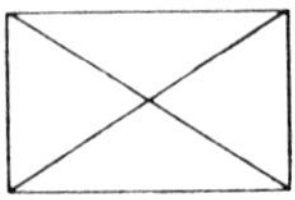

2. 6
3. 4¼
4. 42 chocolates
5. 90
6. 6
7. 4·4
8. 1,000
9. £3·75
10. £2·61
11. 4 boys
12. Sport
13. 200 cm
14. 900 cm.
15.

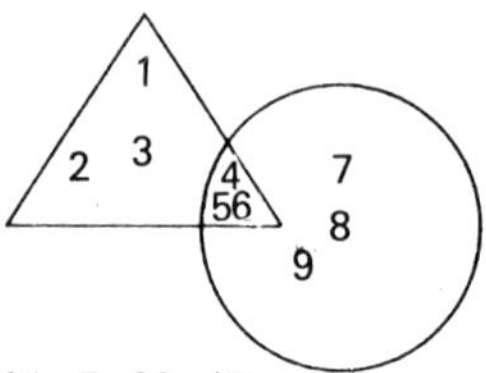

16. 7, 23, 47.
17. +
18. ÷
19. ×
20. 3
21. ÷
22. $\begin{array}{r} 476 \\ +987 \\ \hline 1,463 \end{array}$
23. $\begin{array}{r} 4,678 \\ +3,599 \\ \hline 8,277 \end{array}$
24. $\begin{array}{r} 739 \\ -628 \\ \hline 111 \end{array}$
25. $\begin{array}{r} 731 \\ -472 \\ \hline 259 \end{array}$
26. ½
27. 3·3
28. 0·351
29. 13·888
30. 31·4
31. 2·9
32. 24·2 km.
33. 6,050 m.
34. M, A, X.
35. 22½p
36. 10p, 10p, 2p, ½p
37. 50
38. 15
39. M
40. △
41. 4
42. 7
43. 9
44. 19 cm.^{2}
45. 18 cm.^{2}

46. 19 cm.^{2}
47. 54
48. 3–4 p.m.
49. 14
50. afternoon
51. 9
52. 85 cm.
53. 220 km.
54. 200 km.
55. 279 km.
56. 4 hours
57. 11
58. 15
59. 1,476
60. 351
61. 1,770
62. 90°
63. 45°
64. 0°
65. 315°

Mathematics Test 5

1. Mr Jones, Tom.
   Mrs Jones, Freda, Helen.
   or Mr Jones, Mrs Jones.
   Tom, Freda, Helen.
2. 0·31
3. 20
4. 65
5. 45·3
6. 390p
7. £3.10
8. 41½ l.
9. 18,000
10. £44,500
11. 7·25
12. 9·30
13. 63 min.
14. 14·6
15. £10·35
16. ÷
17. +
18. −
19. 25%
20. 2½ hours
21. 32 boxes
22. 1 and 6, 2 and 7
23. 60p
24. Classes 4, 5, and 8
25. £5·70
26. 160·4 kg.
27. 40·1 kg.
28. 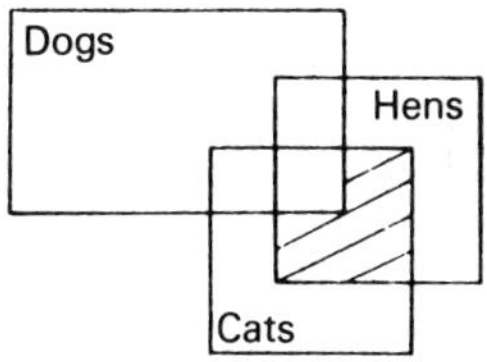

29. 52 km./h.
30. 29 cm.
31. 12·56 m.

32.				13
33. 34.	5			8
35.			6	
36.			15	

37. 60°
38. 6
39. 5
40. 23 143 107
41. 78%
42. 0·67
43. 4·9, 4·6, 4·3, 4·0, 3·7.
44. 14·08
45. 2^5

Intelligence Test 1

1. 6
2. O
3. B
4. word
5. L
6. R
7. firework
8. cod
9. ostrich
10. canary
11. air
12. primrose
13. snow white
14. dog puppy
15. tile roof
16. uncle aunt
17. wide narrow
18. sleek rumpled
19. many few
20. huge tiny
21. 5 8 2
22. 5 6
23. 13 16
24. 13 18
25. 15 10
26. 6 5
27. 27 35
28. 16 19
29. bigger
30. four
31. warm
32. May
33. middle
34. sort
35. walk
36. bus
37. white
38. low
39. south
40. front
41. submarine
42. mean
43. buy
44. family
45. tree
46. water
47. cereal
48. fish
49. shop-keeper
50. fibre

Intelligence Test 2

1. U
2. L
3. take
4. R
5. AA
6. R
7. Q
8. gain
9. fade
10. D
11. L
12. A
13. second week
14. April August
15. stand run
16. $^1/_2$ $^{10}/_{12}$
17. private general
18. ABC EBG
19. 0·00001 0·1
20. small
21. go
22. narrow
23. red
24. niece
25. smooth or gentle
26. peach
27. trawler
28. calf
29. principal
30. diminish
31. little big
32. noisy silent
33. float sink
34. deliver collect
35. happy sad
36. wealthy poor
37. dilute concentrate
38. 9 10
39. 11 8
40. 57 67
41. $^5/_{50}$ $^4/_{60}$
42. 17 19
43. 4 8
44. dry
45. a head
46. high
47. blank
48. knife
49. Ann
50. 8

Intelligence Test 3

1. 1 7 2
2. 10 7 2 3
3. 9 3 5 12 3 2
4. 2 3 7 5 12
5. cat or act
6. penny
7. start or tarts
8. table or bleat
9. A

10. G
11. D
12. C
13. Easter Christmas Day
14. cat man
15. swing stand
16. A 6
17. 3 7
18. DAN
19. THIRTEEN
20. FULFIL
21. drifter
22. railway
23. June
24. cat
25. accuse
26. 3 2
27. 10 12
28. 5 20
29. 4 4
30. 9
31. 9
32. 12
33. 6 and 7 or 8 and 5
34. 13
35. Ann
36. Mary and Ann
37. large small
38. trot gallop
39. rain lake
40. agree deny
41. flat sloping level
42. begin finish conclude
43. ugly beautiful plain
44. brave cowardly courageous
45. never often frequently
46. aid
47. fix or tie
48. fold
49. omitted or dropped or ignored
50. ascend

Intelligence Test 4

1. crowd
2. garden
3. book
4. factory
5. village
6. coat
7. evening
8. waves
9. bridge
10. roof
11. chimney-pot
12. mountain
13. dwelling
14. L
15. Mary
16. Tom
17. Ted
18. June
19. Ted
20. Jill
21. Joan
22. calf cow
23. cage zoo
24. pupil class
25. 15 45
26. lift lower
27. ran
28. down or window
29. was
30. were
31. 10 13
32. 6 18
33. 10 30
34. 58 69
35. 7·3 73,000
36. Mary
37. Dora
38. Vera
39. Vera
40. Jill and Maud
41. red
42. mg.
43. many
44. clumsy
45. clear
46. assist
47. contract
48. U
49. C
50. D

Intelligence Test 5

1. + +
2. − +
3. − +
4. + −
5. + +
6. Jane
7. Sally
8. Mary
9. Mary
10. Jane
11. m. cm. mm.
12. ash elm oak
13. pond stream sea
14. cycle bus car
15. hat helmet cap
16. Fred
17. John
18. John
19. John
20. Robert
21. none
22. shoe
23. day
24. den
25. pond
26. 20 40
27. 6 8
28. 10 15
29. 6 3
30. 4 16
31. bird air
32. beginning end
33. bell bicycle
34. laundered soiled
35. hum dynamo
36. eleventh
37. brown
38. bird
39. walk
40. banana
41. 5
42. 6
43. 37
44. 25
45. 9
46. 5
47. 11
48. 1C
49. 12
50. 64

Intelligence Test 6

1. (*b*)
2. (*c*)
3. (*d*)
4. (*b*)
5. (*c*)
6. (*c*)
7. (*b*)
8. (*c*)
9. (*a*)
10. (*a*)
11. (*b*)
12. (*a*)
13. (*c*)
14. (*c*)
15. (*d*)
16. (*d*) (*c*)
17. (*a*) (*b*)
18. (*b*) (*a*)
19. (*c*) (*d*)
20. (*b*) (*c*)
21. MM OO
22. IJ KL
23. AZ BY
24. MX NX
25. JKL LMN
26. Mary
27. Tom
28. Arithmetic
29. Tom
30. Frank and Tom
31. 5
32. 6
33. 2
34. Sunday
35. dull
36. 9
37. 2
38. 9
39. K
40. 1
41. CAM
42. EAR
43. DAME
44. BIND
45. MEND
46. 9
47. 1
48. 8
49. 4
50. 4

Intelligence Test 7

1. R
2. N
3. A
4. NOISUTNOC
5. E
6. N
7. I
8. T
9. animal
10. weapon
11. water
12. sound
13. measure
14. transport
15. home
16. emblem
17. OP PQ
18. DO EO
19. SS RR
20. VW XY
21. LK JI
22. DEF EFG
23. 13 12
24. 21 23
25. 18 24
26. $5/8$ $6/9$
27. 6 8
28. 167 178
29. 25 29
30. last first
31. wood desk
32. ceiling floor
33. hen egg
34. sugar vinegar
35. captain crew
36. AWAKE
37. DARK
38. QUIET
39. MEND
40. LOWER
41. KNPS
42. YPL
43. PBXPOXBCZ
44. XASVBXAM
45. hour
46. village
47. 648
48. shop
49. wind
50. destroyer
51. 1
52. 5
53. 4
54. 4
55. 13
56. M N
57. O
58. long short
59. river tree
60. snow bed
61. cushion pillow
62. scale feather
63. airport aeroplane
64. find hide
65. girl boy
66. gain endeavour
67. boy class
68. game
69. size
70. engine
71. herd
72. subject
73. number
74. newspaper
75. year
76. street
77. frequency
78. sheep
79. C
80. write

81. field
82. needle
83. cabbage
84. rugged
85. say
86. sow
87. huge tiny
88. forbid allow
89. arrive depart
90. confirm deny
91. please annoy
92. permit prevent
93. tighten relax
94. follow precede
95. boisterous tranquil
96. release arrest
97. baby grandfather
98. cycle aeroplane
99. white black
100. 999 g. 3 kg.

Intelligence Test 8

1. B
2. C
3. D
4. C
5. D
6. D
7. C
8. D
9. A
10. C
11. C
12. D
13. C
14. A
15. C
16. B
17. C
18. B
19. A
20. D
21. A
22. C
23. D
24. B
25. B
26. C
27. B
28. C
29. D
30. B
31. D
32. E
33. B
34. C
35. E
36. C
37. E
38. B
39. E
40. E
41. E
42. B
43. B
44. E
45. A
46. A
47. B
48. A
49. A
50. D